Thrive in Bioscience | Revision Guides

Other titles in the Thrive in Bioscience series

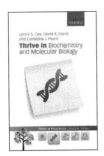

Thrive in Biochemistry and Molecular Biology

Lynne S. Cox, David A. Harris, and Catherine J. Pears

Thrive in Genetics

Alison Thomas

Thrive in Ecology and Evolution

Alan Beeby and Ralph Beeby

Thrive in Cell Biology

Qiuyu Wang, Chris Smith, and Emma Davis

Forthcoming: **Thrive in** Immunology

Anne Cunningham

Thrive in Human Physiology

Ian Kay
Associate Head of School, School of Healthcare Science,
Manchester Metropolitan University

Gethin Evans
Senior Lecturer, School of Healthcare Science,
Manchester Metropolitan University

Thrive in Bioscience | Revision Guides

OXFORD
UNIVERSITY PRESS

OXFORD
UNIVERSITY PRESS

Great Clarendon Street, Oxford, OX2 6DP,
United Kingdom

Oxford University Press is a department of the University of Oxford.
It furthers the University's objective of excellence in research, scholarship,
and education by publishing worldwide. Oxford is a registered trade mark of
Oxford University Press in the UK and in certain other countries

© Ian Kay and Gethin Evans 2014

The moral rights of the authors have been asserted

Impression: 7

Published in the United States of America by Oxford University Press
198 Madison Avenue, New York, NY 10016, United States of America

British Library Cataloguing in Publication Data

Data available

Library of Congress Control Number: 2013954696

ISBN 978–0–19–966248–7

Printed and bound by
CPI Group (UK) Ltd, Croydon, CR0 4YY

Links to third party websites are provided by Oxford in good faith and
for information only. Oxford disclaims any responsibility for the materials
contained in any third party website referenced in this work.

Contents

Contents

Four steps to exam success

1 Review the facts

This book is designed to help you learn quickly and effectively:

- Information is set out in bullet points, making it easy to digest
- Clear, uncluttered illustrations illuminate what is said in the text
- Key concept panels indicate the essential learning points for a topic

2 Check your understanding

- Try the questions in each chapter and online multiple-choice questions to reinforce your learning
- Download the flashcard app to master the essential terms and phrases

3 Take note of extra advice

- Look out for hints for getting those precious extra marks in exams

4 Go the extra mile

- Explore other sources of information—including human physiology textbooks—to take your knowledge and understanding one step further.

Go to the Online Resource Centre for more resources to support your learning, including:

- Online quizzes, with feedback
- A flashcard glossary, to help you master the essential terminology

www.oxfordtextbooks.co.uk/orc/thrive/

online resource centre

1 Introduction

The *Oxford English Dictionary* defines physiology as 'The branch of science that deals with the normal functioning of living organisms and their systems and organs. Also: the functional processes of an organism, organ, or system'. All organisms—microbes, plants and animals—have their own unique physiology, which can be studied at a variety of levels of organization from the molecular to the organ level. This book concerns itself with the normal function of the human body. Though primarily organ based, it will consider the underlying molecular and cellular elements, which describe 'how we work'. Intimately linked with physiology is an understanding of the structures within the body—the link between structure and function in physiology cannot be overemphasized.

Key concepts

- Physiology is the study of how the human body works.
- In terms of structure, the body shows a hierarchical nature of organization from the molecular through to the organismal level.
- In terms of elemental composition, the vast majority of the body is formed from hydrogen, oxygen, carbon, and nitrogen.
- In terms of body composition, the majority of an individual's body weight is accounted for by water, which is found in a variety of compartments.

continued

- Water within the body has a variety of solutes dissolved in it and thus forms extracellular fluid and intracellular fluid, each of which has a unique composition.
- Homeostasis is the maintenance of a relatively stable internal environment.
- In terms of maintaining homeostasis, the most important control mechanisms are achieved through negative feedback control.

1.1 STRUCTURE OF THE BODY

It is possible to consider the structure of the body at a number of different levels, starting with the molecular level and building up to the organismal (whole body) level.

Molecular level

- At the molecular level, the body consists of four macromolecules: proteins, carbohydrates, lipids, and nucleic acids.
- **Proteins** are polymers of amino acids—the body utilizes some 20 different amino acids from which it synthesizes proteins (Figure 1.1). Given that a typical protein may contain several hundred amino acids, the number of permutations in which the amino acids may be arranged is vast—ultimately, proteins are unique molecules.
- Proteins have a variety of roles within the body, they act as **receptors** for **hormones** and **neurotransmitters**; they act as enzymes and they act as carrier molecules transporting material across cell membranes.
- **Carbohydrates** exist in many forms—monosaccharides (e.g. glucose) or polysaccharides, which are polymers of many monosaccharides (e.g. glycogen) (Figure 1.2).
- The primary role of carbohydrates is to act as an energy source for the multitude of reactions occurring within the body. Some organs (e.g. the brain) have an absolute dependence on glucose as a source of energy.
- Polysaccharides function as an energy store, i.e. glycogen can be broken down into its constituent monomers (glucose) to provide a source of energy.

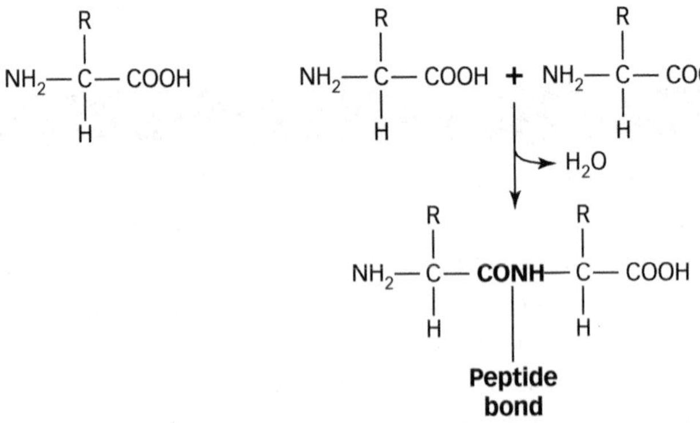

Figure 1.1 The structure of a typical amino acid and its polymerization to form peptides and proteins

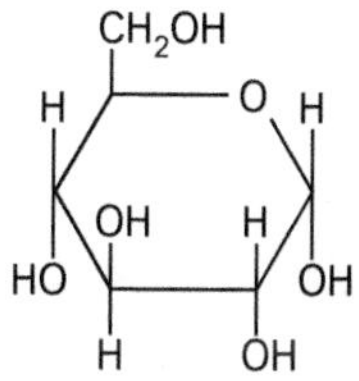

$$CH_2OH$$

Figure 1.2 The structure of glucose

- Carbohydrates may combine with proteins and lipids to form glycoproteins and glycolipids, respectively.
- A final role for one group of carbohydrates is to contribute to the structure of DNA and RNA.
- **Lipids** are water-insoluble compounds, unlike proteins and carbohydrates, which perform a variety of roles in the body. Lipids are an important component in the formation of cell membranes; they act as an energy source and some of them (e.g. the prostaglandins and steroids) act as hormones.
- **Nucleic acids**—DNA and RNA—are the principal components in the storage and transfer of genetic information. DNA is formed from a sugar molecule (ribose) linked to a variety of bases (thymine, cytosine, adenine, and guanine).
- A molecule of sugar combined with a base produces a structure called a nucleoside. When this becomes linked to a phosphate group it forms a nucleotide. DNA (and RNA) can be thought of as polymers of nucleotides (Figure 1.3).
- RNA differs from DNA in that its sugar component is ribose (instead of deoxyribose) and uracil replaces thymine as a base.
- DNA is stored, in conjunction with histone proteins, in the nucleus of the cell in the form of chromosomes.

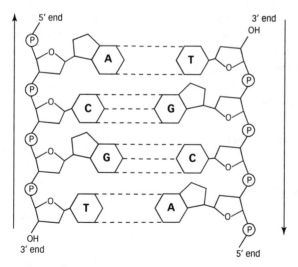

Figure 1.3 The structure of DNA

Structure of the body

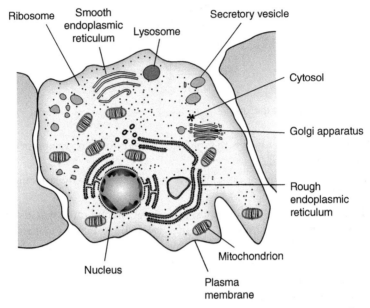

Figure 1.4 The structure of a typical animal cell showing the principal organelles

Cellular level

- Cells are the basic building blocks of the human body. Whilst displaying tremendous diversity, they all share common characteristics. A 'typical' cell is shown in Figure 1.4.
- All cells are bound by a **plasma membrane**. The membrane consists of lipids and proteins, and works to regulate the passage of material into and out of the cell. Some substances can move across the membrane by simple diffusion (depending on their size and lipid solubility), whilst others need specialized energy-consuming transport molecules. The structure of the membrane is shown in Figure 1.5.
- Each **organelle** within a cell is also bounded by a membrane. This means that there is a high degree of compartmentalization within each cell, ensuring that all functions are separated from each other.

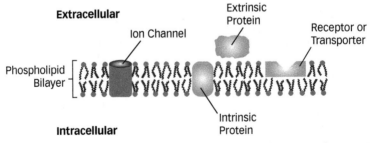

Figure 1.5 The structure and principal components of the plasma membrane

- The **nucleus** lies within the cytoplasm of the cell, and at its boundary is the nuclear membrane. The nucleus contains the DNA of the cell which, in conjunction with histone proteins, forms chromosomes.
- **Mitochondria** within the cytoplasm act as energy transducers. They do this by oxidizing glucose and fatty acids and in doing so produce adenosine triphosphate (ATP). ATP is the energy source on which cells depend. The more active a cell is (e.g. a muscle cell), the greater the number of mitochondria in it.
- Running throughout the cytoplasm is a series of membrane-bound organelles called the **endoplasmic reticulum**. There are two types: rough and smooth.
- Rough endoplasmic reticulum has its appearance because it is associated with ribosomes. Ribosomes are the site of protein synthesis within the cell.
- Smooth endoplasmic reticulum is associated with the production of lipids. It also acts as a reservoir for calcium. Calcium is an important intracellular signalling molecule; when it is released it can trigger physiological processes within the cell (e.g. contraction in the case of muscle cells).
- The **Golgi apparatus** is a structure which is involved in the modification and packaging of proteins after they have been synthesized and prior to their release.
- **Lysosomes** are membrane-bound vesicles found in the cytoplasm. They contain a variety of hydrolytic enzymes. These are enzymes which are responsible for degradation and therefore contribute to the 'recycling' of, for example, components of the membrane.
- Within the cell there are a series of structures known as the **cytoskeleton**, which can be thought of as internal scaffolding and gives a cell its 3D structure. The cytoskeleton is made up of three components: actin filaments, intermediate filaments, and microtubules.

Looking for extra marks?

Mitochondria contain their own DNA and as such undergo mitotic cell division. The DNA found inside mitochondria comes solely from the mother, unlike nuclear DNA which comes from both parents

Tissue level

- Individual cells aggregate to form tissues.
- There are four main types of tissue:
  - epithelial tissue
  - connective tissue
  - neural tissue
  - muscular tissue.

  It is important to realize that each tissue type has many subtypes.
- **Epithelial tissue** is covering tissue—this may be the skin which protects us from the outside world or tissue which lines the gastrointestinal tract and airways.

Epithelia may take on specialized roles; for example, they may form glands such as sweat glands. Epithelia are generally located on top of a layer of connective tissue—this provides structure and support.

- **Connective tissue** acts to support other tissues, and to link tissues and cells together. The most common form of connective tissue is loose connective tissue. Within this are elastic fibres and collagen fibres—the latter provide strength and flexibility. Other examples of connective tissue are bone, cartilage, tendons, ligaments, adipose tissue, and blood.
- **Muscular tissue** is specialized tissue that has the ability to contract. Therefore its principal role is that of movement. There are several types of muscular tissue: striated muscle, smooth muscle, and cardiac muscle. These will be discussed in more detail in Chapter 4. In humans, striated muscle also plays a role in the generation of heat. This is of importance when body temperature drops, and one of the responses to this is shivering.
- **Neural tissue** is composed of neurons and glial cells. Neurons are the functional building blocks of the nervous system. Glial cells ensure that neurons are able to work optimally.

Looking for extra marks?

Some epithelial cells demonstrate polarity. The aspect of the cell which faces the outside world (e.g. the lumen of the gastrointestinal tract) is called the apical surface. The opposite face to this, i.e. lying on connective tissue, is called the basolateral surface.

Loose connective tissue, as described above, contains two important cell types. A group of cells called fibroblasts secrete the proteins that are found. A second group of cells called macrophages have phagocytic-like activity and are responsible for the removal of debris and dead cells.

Organ/system level

- Several different tissue types combine to form organs, and ultimately organs work cooperatively to form organ systems. For example, the cardiovascular system consists of both the heart and the vascular system.
- There are a number of 'systems' within the body:
  - integumentary
  - cardiovascular
  - lymphatic
  - respiratory
  - endocrine
  - nervous
  - reproductive
  - renal

 o musculoskeletal

 o digestive.

- Although the body consists of these individual systems, it is important to realize that it is their integrative nature which needs to be taken into account when considering physiology as a whole.

1.2 COMPOSITION OF THE BODY

- At an elemental level, carbon, oxygen, hydrogen, and nitrogen make up some 97% of the human body, with the remainder being minerals such as sodium, potassium, calcium, etc.
- Approximately 60% of the total body weight of an average person (in physiology, this is a young male weighing 70kg) is water. Females generally have a higher proportion of body fat, so this figure drops to 50–55%.
- Water is an absolute requirement for life. Water within the body exists in different compartments.
- It is possible to divide total body water (TBW) into that inside cells— **intracellular fluid (ICF)**—and that outside cells—**extracellular fluid (ECF)**. It is termed fluid rather than water because it has solutes dissolved in it.
- ECF may be further subdivided into blood plasma, interstitial fluid, and transcellular fluid.
- Blood plasma is the liquid component of blood and is the fluid medium in which blood cells are suspended. Plasma accounts for about 8% of TBW.
- **Interstitial fluid** is the fluid which surrounds each and every cell, and accounts for about 25% of TBW.
- **Transcellular fluid** is the fluid found in spaces such as the eye, the ventricles of the brain, synovial joints, etc. It accounts for about 2% of TBW.
- The composition of ECF and ICF is quite different—this is achieved by the fact that they are separated by plasma membrane. The plasma membrane displays a differential permeability to different substances. The membrane also contains transport mechanisms, which are capable of transporting substances across it, and selective ion channels, which allow the passage of ions from one side of the membrane to the other (Figure 1.5).
- The typical composition of ICF and ECF is shown in Table 1.1.

Substance	ICF (mM)	ECF (mM)
Na^+	20	145
K^+	150	4
Cl^-	4	115
HCO_3^-	10	30
Ca^{2+}	0.1×10^{-3}	2

Table 1.1 The composition of ICF and ECF

- Essentially, ECF is a Na^+-rich solution, whilst ICF is a K^+-rich solution. ICF also contains significant quantities of protein.
- Although these solutions are relatively constant, there is a continual movement of substances, but in the long term the composition remains stable.
- The composition of transcellular fluids (e.g. cerebrospinal fluid) may differ significantly from that of ECF. This indicates that these fluids are not simply produced by filtration of blood—substances are added to or taken away from plasma to produce them.

Looking for extra marks?

The unequal composition of ICF and ECF can be 'used' to influence the transport of materials across cell membranes. For example, the active transport of glucose from the lumen of the gut into the epithelial cells of the gut (enterocytes) is dependent upon Na^+ ions. Both glucose and Na^+ bind to the transport molecule, bringing both into the cell. Glucose is then transported from the enterocyte into plasma whilst the Na^+ is transported out of the cell via the activity of the Na^+/K^+-ATPase transporter. This is a separate transport molecule which transports Na^+ out of a cell in exchange for K^+—this helps to maintain the correct ionic composition of ICF and ECF.

1.3 HOMEOSTASIS

- The literal translation of **homeostasis** is 'steady state'.
- In terms of physiology, this means the maintenance of a relatively constant internal environment—composition of fluids, temperature, hormone levels, etc. These variables may fluctuate in the short term, but in the long term they remain relatively constant.
- Each variable regulated by homeostasis has a set point (e.g. body temperature) has a set point of 37°C. However, in reality, each variable has a working range. For example, although the set point of body temperature may be 37°C, the working range may be 36.5–37.5°C.
- There are five components of homeostatic control systems:
  - **detectors** are structures which monitor variables (e.g. thermoreceptors monitor temperature, baroreceptors monitor pressure in the cardiovascular system, and so on).
  - **afferent neurons**
  - **comparator**—this 'knows' what the set point should be
  - **efferent neurons**
  - **effectors** are structures which produce a response (e.g. muscle contraction).

  These components are found in all homeostatic control systems.
- These components are arranged such that control is achieved by the process of **negative feedback**—when a variable is disturbed, changes are initiated which return it to its set point. This can be illustrated with an example—regulation of blood pressure.

- o The set point for 'normal' blood pressure is 120/80mmHg.
- o Let us assume that blood pressure drops—this drop is detected by specialized receptors in the walls of some arteries called baroreceptors.
- o The baroreceptors signal this drop back to the brainstem region of the brain via transmission of action potentials.
- o In the brainstem, this signal is compared with the normal or desired value. In this case there is a mismatch between the two and an error signal is generated.
- o This error signal is conveyed via efferent neurons to effectors.
- o In this case the effectors are the heart, to increase the force and rate of contraction, and the smooth muscle surrounding arteries, which contracts. The overall consequence of activity in these structures is that blood pressure increases and returns to its set point of 120/80mmHg.
- o As blood pressure returns to its set point, activity in the baroreceptors decreases and consequently the processes which have been activated reduce their activity. Figure 1.6 illustrates this mechanism.

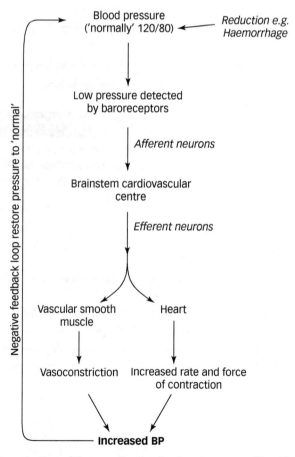

Figure 1.6 Organization of the negative feedback system controlling blood pressure

The role of negative feedback in the maintenance of homeostasis cannot be overemphasized.

- Whilst negative feedback is important, it can only correct changes after they have occurred. It would be more economical if changes could be minimized, therefore reducing the magnitude of a negative feedback response.
- **Feedforward** is a process that limits changes in physiological variables, thus limiting the response required to bring them back to their set point.
- A good example of this is wearing clothing appropriate to the weather. For example, in cold weather, coats, hats, and scarves are worn. This reduces any drop in body temperature that would be seen in a person who was wearing shorts and a T-shirt. Since the drop in body temperature is reduced, fewer physiological responses need to be initiated to maintain body temperature at its set point.
- The opposite of negative feedback is positive feedback. In this case, a change in a regulated variable initiates responses which make it change even more.
- For example, if blood pressure was subject to positive feedback, a drop in blood pressure would initiate responses that would make it drop even further, which would then initiate responses to drop further, and so on.
- There are some examples of positive feedback in physiology. They include:
  ○ the process of birth (parturition)
  ○ sodium ion entry during the depolarizing phase of the action potential
  ○ the role of luteinizing hormone in the process of ovulation.

Looking for extra marks?

Although physiological variables have a working range, it is possible for them to be reset. A good example of this is the resetting in the hypothalamus of the body temperature during infections. This results in fever, which helps to fight off the infection.

Check your understanding

Describe the organization of the human body. (*Hint: consider the different levels of organization from molecular through to organs*)

What are the main components a homeostatic control system? (*Hint: think about how the body detects change and responds to such change*)

Why is negative feedback the main control system in the body? (*Hint: consider what would happen to physiological processes and homeostasis if positive feedback prevailed*)

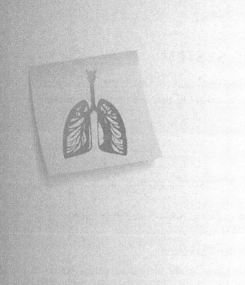

2 The nervous system

The fundamental role of the nervous system is control and coordination (in conjunction with the endocrine system—see Chapter 3). The functional elements of the nervous system are neurons, which are supported by a second group of cells, glial cells, whose overarching role is to support neurons and ensure that they function effectively. Neural control is characterized by its electrical mediation, its rapidity, and its duration of response.

Key concepts

- The nervous system consists of two cell types—glial cells and neurons.
- Neurons are characterized by a large resting membrane potential, the polarity of which is reversed during an action potential.
- Neurons communicate with each other by the release of chemical compounds called neurotransmitters.
- There are a number of ways of describing the organization of the nervous system: central/peripheral and efferent/afferent.
- The brain represents the highest level of control in the nervous system.
- The spinal cord is capable of undertaking relatively complex neural responses.
- The autonomic nervous system is an involuntary part of the nervous system that regulates smooth and cardiac muscle.

2.1 CELLS OF THE NERVOUS SYSTEM

There are two cell types in the nervous system—glial cells (of which there are several types) and neurons. Neurons are the functional cells, but they are supported in their role by glial cells.

Glial cells

- **Glial cells** constitute approximately 90% of the cellular elements of the nervous system.
- They retain the ability to undergo mitotic cell divisions—hence the majority of brain tumours are gliomas (of which there are several types).
- Several types of glial cell are found in the nervous system.
  - **Astrocytes** are star-shaped cells that are found in the central nervous system (CNS). These cells are important in establishing the **blood–brain barrier** (**BBB**)—this isolates the brain from the blood, which may contain substances that may be damaging to neurons.
  - **Microglia** are phagocytic cells, which remove, for example, damaged neurons and microorganisms. They form part of an active immune system—this is particularly important as antibodies are unable to cross the BBB.
  - **Oligodendrocytes** form the myelin sheath of CNS neurons. In the peripheral nervous system (PNS), this role is performed by Schwann cells.
  - **Ependyma** are glial cells which line the ventricles of the brain and form cerebrospinal fluid (CSF).

Looking for extra marks?

In recent years, glial cells have been found to play a more dynamic role in the nervous system than previously thought. It is now known that some of them have the ability to synthesize and release neurotransmitters and therefore influence the activity of neurons. Equally, some of them may release cytokines, which influence neuronal survival and synapse formation.

Neurons

- **Neurons** constitute the functional cellular elements of the nervous system in that they are capable of generating and transmitting action potentials.
- Neurons consist of three functional regions: dendritic (input from other neurons or sensory receptors); somatic (cell body typical of all cells); axonic (output to other neurons or effectors, e.g. muscles and glands).
- It is possible to classify neurons either using structural or functional criteria.
- Structural classification, based on the projections originating from the cell body, gives three groups (Figure 2.1):

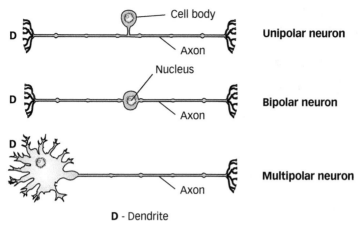

Figure 2.1 The structural classification of neurons

- ○ unipolar—one projection (e.g. peripheral sensory neurons)
- ○ bipolar—two projections (e.g. retinal neurons)
- ○ multipolar—many projections (e.g. motor neurons).
- Functional classification, based on the direction of action potential movement relative to the CNS similarly gives three groups (Figure 2.2):
  - ○ afferent (sensory) neurons—action potentials travelling towards the CNS
  - ○ efferent (motor) neurons—action potentials travelling away from the CNS
  - ○ **interneurons**—action potentials travelling within the CNS.
- A typical neuron is shown in Figure 2.3. The highly extensive dendritic branching allows inputs from other neurons. A typical neuron may receive up to 10000 inputs.
- An individual neuron integrates the inputs it receives and, if appropriate, an action potential is generated and transmitted along the axon. The axon may be up to 1m in length. The presence of a **myelin sheath**, formed from Schwann cells, increases the conduction velocity of the action potential.
- Axons terminate at axon terminals, which form the presynaptic regions of a synapse and allow communication with other neurons or structures.

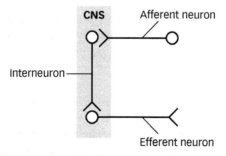

Figure 2.2 The functional classification of neurons

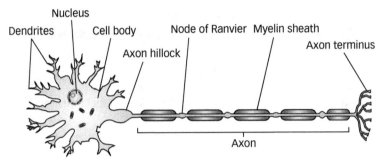

Figure 2.3 The structure of a typical neuron

2.2 NEURONAL FUNCTION

Neurons function by virtue of the fact that they have a potential difference (voltage) across their cell membrane and the polarity of this potential can transiently change to signal activity. However, although they are fundamentally electrical in nature, they communicate with each other by releasing specialized molecules called neurotransmitters.

Resting membrane potential

- It is possible to measure a potential difference (voltage) across the membrane of an individual neuron. This is done by placing one microelectrode in the cytoplasm of the cell whilst leaving a second one on the outside of the cell.
- Since this potential difference is recorded whilst the neuron is at rest, it is called the **resting membrane potential** (**RMP**). It is of the order of -80mV, with the inside being negatively charged with respect to the outside.
- The RMP is established by three factors:
  - a difference between the composition of fluids within the neuron (intracellular fluid (ICF)) and that outside the neuron (extracellular fluid (ECF)).
  - differential permeability of the neuronal membrane to different ions found in the ICF and the ECF.
  - the presence of transport proteins in the membrane which are capable of moving ions across the membrane.
- The ICF is rich in K^+ ions and poor in Na^+ ions—concentrations are approximately 140mM and 5mM, respectively. In the ECF, these concentrations are essentially reversed. However, note that, through the presence of other ions in both the ICF and ECF, the 'numbers' of positive and negative ions in the two compartments are equal.
- At rest, the neuronal membrane is permeable to K^+ ions—the presence of the so-called K^+ leak channels ensures that K^+ ions are free to move down their outwardly acting concentration gradient from the ICF to the ECF.
- As K^+ ions move they take a positive charge with them and leave behind their corresponding associated negative charge. As this happens, a potential difference begins to be established.

- As K$^+$ ions move out, it becomes increasingly difficult for subsequent ions to move out because of the development of an inwardly acting electrical gradient.
- At some point the outwardly acting concentration gradient is equal and opposite to the inwardly acting electrical gradient—this is called the K$^+$ electrochemical equilibrium. At this point, a potential difference has been established—the K$^+$ electrochemical equilibrium potential. This value approximates to, but is not identical to, the measured RMP.
- The principal extracellular ion is Na$^+$. There is an inwardly acting Na$^+$ concentration gradient. However, membrane permeability to Na$^+$ is only a hundredth of that to K$^+$. Therefore there is a small movement of Na$^+$ ions which brings a positive charge back into the neuron. This acts to reduce the value of the potential difference generated by the movement of K$^+$ alone.
- Overall, then, the RMP is established primarily by the efflux of K$^+$ ions against a small influx of Na$^+$ ions.
- Over time, the neuron loses K$^+$ and gains Na$^+$. Within the membrane, there is a **Na$^+$/K$^+$-ATPase pump** which transports three Na$^+$ ions out in exchange for two K$^+$ ions in. This restores the concentration gradients of these ions, upon which the RMP depends. Since there is an unequal exchange of ions, this pump also makes a small contribution to the RMP itself.

Looking for extra marks?

The final value of the RMP in any given neuron is the sum of all ions in both the ECF and the ICF and their relative permeabilities. This can be predicted by the Goldman constant field equation.

Action potentials

- **Action potentials (APs)** are the 'currency' of the nervous system.
- APs are transient reversals of the normal polarity of the RMP, i.e. the inside of the neuron changes from being negatively charged to positively charged and back to being negatively charged all within 2ms. A typical intracellular recording of an AP is shown in Figure 2.4.
- APs may be initiated by activity in other neurons, by stimulation of sensory receptors, or experimentally by the application of an electrical or drug-induced stimulus.
- The AP can be divided into three phases: **depolarization**, **repolarization**, and **after-hyperpolarization**.
- The depolarization phase takes the interior of the neuron from an RMP of approximately −80mV to a value of +30mV. The mechanism responsible for this is the rapid opening of membrane Na$^+$ channels and the influx of Na$^+$ ions.
- At the peak of the AP, the Na$^+$ channels close and a series of K$^+$ channels open. Consequently, K$^+$ ions, and thus positive charge, leave the neuron. Therefore the membrane potential begins to decrease—this is the repolarizing phase.
- Compared with the Na$^+$ channels, the K$^+$ channels, which open during repolarization, are slow to close. Consequently, the loss of K$^+$ ions exceeds that

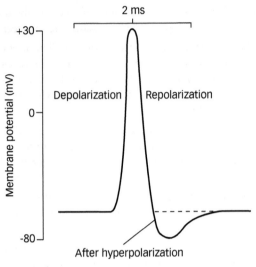

Figure 2.4 Intracellular recording of an action potential

necessary to return the neuron to the RMP. Therefore there is a temporary drop in membrane potential below the RMP—this is the after-hyperpolarization phase.

- At the end of the AP, a neuron has gained Na$^+$ ions and lost K$^+$ ions. The activity of the Na$^+$/K$^+$-ATPase pump restores these ions to their original pre-AP location and the overall membrane potential is returned to the RMP.

- Not all stimuli will be of sufficient intensity to generate an AP. Though they may be sufficient to generate small depolarizations, in order to generate an AP, the membrane potential must reach a critical value—the **threshold**.

- The threshold is generally about 15mV above the RMP. If the RMP reaches this value, the Na$^+$ channels open instantaneously and depolarization occurs.

- The **refractory period** is a period of time during which the neuron is either unresponsive to a second stimulus or an increased stimulus intensity is required to generate a second AP.

- The absolute refractory period is the time during which a neuron is unresponsive to a second stimulus, irrespective of intensity. This period coincides with the depolarizing phase moving into repolarization of the AP. It corresponds to the period when the Na$^+$ channels are inactivated.

- Inactivation of Na$^+$ channels follows their rapid opening and closure during depolarization. From being inactivated they become closed and are then able to open again.

- The relative refractory period corresponds to the depolarizing and after-hyperpolarizing phase of the AP. During this phase a second stimulus, of greater magnitude than that which generated the first AP, may generate a second AP.

- APs generated at the axon hillock are transmitted along the entire length of the axon by the generation of local currents (Figure 2.5).

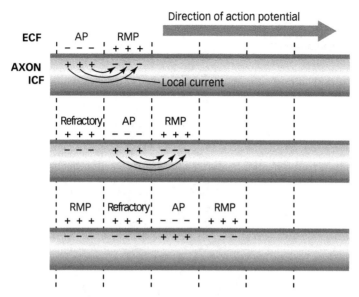

Figure 2.5 Saltatory conduction of an action potential along an axon

- Regions of depolarized membrane depolarize the adjacent region, which depolarizes the next, and so on.
- The presence of the relative refractory period ensures that the movement of the AP is unidirectional.
- Two important factors which influence the speed of conduction of the AP are the diameter of the axon—the larger the faster—and the presence of a myelin sheath. In myelinated neurons, depolarizations only occur at the **nodes of Ranvier**; therefore conduction velocity is increased.
- Conduction velocity ranges from ~2.5m/s in small-diameter non-myelinated neurons to ~120m/s in large-diameter myelinated neurons.

Looking for extra marks?

The Na$^+$ channels described previously are examples of voltage-gated channels. At rest, i.e. at the RMP, these channels are closed. As a neuron begins to depolarize, these channels open. They can be distinguished from a second set of channels—ligand-gated channels. Ligand-gated channels are characterized by opening in response to ligand binding—either a neurotransmitter or a drug.

Synaptic neurotransmission

- **A synapse** is the gap between an axon terminal and another neuron or structure (e.g. muscle cells). Mechanisms which allow an action potential to 'pass' from the axon terminal to the second structure must be in place.

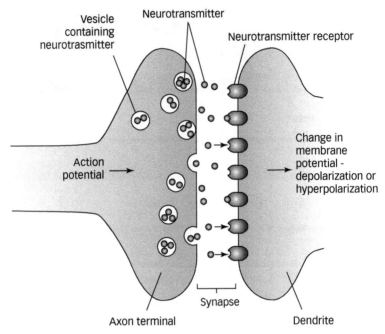

Figure 2.6 Organization of a typical synapse

- Synapses may operate either electrically or chemically—chemical neurotransmission is most common and will be considered here. The organization of a typical neuron–neuron synapse is shown in Figure 2.6.
- Action potentials reach the axon terminal and result in its depolarization. This facilitates the opening of voltage-gated Ca^{2+} channels, which allows entry of Ca^{2+} into the axon terminal.
- The rise in intracellular Ca^{2+} levels activates a variety of intracellular protein kinases (e.g. Ca/calmodulin-dependent protein kinase I).
- Activation of the kinases results in the fusion of the vesicle containing the **neurotransmitter** with the membrane of the axon terminal and release of the neurotransmitter into the synapse.
- The neurotransmitter binds to selective receptors on the post-synaptic membrane and results in a change in membrane potential. These changes may be either excitatory—a depolarizing response caused by the influx of Na^+ ions—or hyperpolarizing—caused by the loss of further K^+ ions or the influx of Cl^- ions.
- Any changes in membrane potential, which are short-lived and decrease in size as they move away from the point of initiation, are transmitted towards the axon hillock. If, by the time they reach the axon hillock, they are sufficient to raise the membrane potential to threshold, an action potential will be initiated in the second neuron.

- A typical neuron may have up to 10,000 synaptic inputs. The neuron will simply integrate and sum these responses at the axon hillock to determine whether an AP will be initiated.
- The final step in the process is the detachment of the neurotransmitter from its receptor followed by metabolism and inactivation. Metabolism may occur in the synapse (extracellular) or it may be intracellular following active uptake of the neurotransmitter into neurons or glial cells.

Looking for extra marks?

The process of synaptic neurotransmission is much more complex than described here. It is now known that there are differences between central and peripheral synapses and that a bewildering array of proteins are involved in the process.

A host of substances act as neurotransmitters, with perhaps the best known being acetylcholine and noradrenaline. However, many others are known to exist. Deficiencies with synaptic neurotransmission account for a variety of neurological diseases. For example, loss of dopamine in the basal ganglia is responsible for Parkinson's disease. A deficiency of the neurotransmitter serotonin is implicated in the pathogenesis of depression.

The fact that neurotransmission is a multistep process allows therapeutic intervention at different levels to offer treatments in disease states. For example, depression can be treated pharmacologically by the use of drugs that prevent the uptake of serotonin from synapses, effectively increasing the amount of serotonin available for synaptic transmission.

2.3 ORGANIZATION OF THE NERVOUS SYSTEM

The organization of the nervous system is shown in Figure 2.7. At its most basic level of organization, the nervous system is composed of the central nervous system (brain and spinal cord) and the peripheral nervous system.

CNS—the brain

- The brain can be conveniently divided into a number of functional regions. It is important to realize that its overall function requires the integration of neural activity across a number of these regions at any given time. A saggital section of the brain is shown in Figure 2.8.
- The brain is enclosed within the bones of the skull and is covered by three further membranes—the **pia mater**, the **arachnoid mater**, and the **dura mater**. There is a space between the pia mater and the arachnoid mater—the subarachnoid space.
- The subarachnoid space contains **cerebrospinal fluid** (CSF). CSF is produced by the ependymal cells, which line the lateral ventricles of the brain. It has a number of roles including that of acting as a 'shock absorber'.

Organization of the nervous system

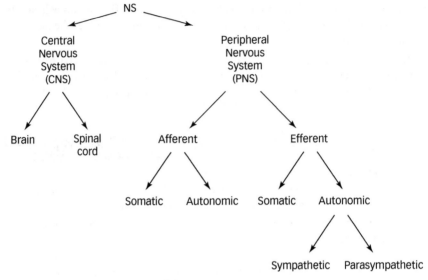

Figure 2.7 General organization of the nervous system

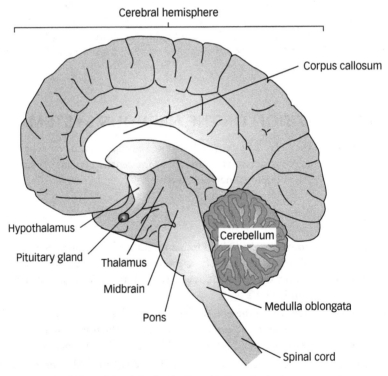

Figure 2.8 A saggital section of the brain showing its principal components

CNS—the brainstem

- The **brainstem** is the most fundamental region of the brain. It is composed of three regions—the **medulla**, the **pons**, and the **midbrain**. All neural activity between the brain and the rest of the body passes through the brainstem.
- The medulla is continuous with the spinal cord.
  - The medulla contains a number of 'centres' (groups of neurons with similar roles) which coordinate cardiovascular and ventilatory activity. For example, it contains the medullary chemoreceptors, which monitor levels of CO_2 and adjust ventilation accordingly. It contains other centres which regulate swallowing and sneezing amongst other activities.
  - The medulla is also the site of origin of five of the twelve cranial nerves. The cranial nerves are, with one exception (the 10th cranial nerve—the vagus), responsible for the innervations (sensory and motor) of structures in the head and neck. These nerves are responsible for a number of cranial reflexes (e.g. the papillary reflex—when light is shone on the eye, the pupil reflexly constricts to regulate the amount of light entering the eye).
  - The medulla also contains structures on its surface called the pyramids. These are triangular-shaped collections of neurons which originate in cortical regions of the brain and terminate in the spinal cord. These neurons are involved in regulating motor activity—they terminate on and influence motor neurons, which control skeletal muscle. This arrangement is termed the corticospinal pathway. Fibres within this pathway decussate near the boundary between the medulla and spinal cord. Decussation means crossing over—hence neurons which have originated in the right motor cortex control the left side of the body and vice versa.
  - There is a second group of cortical neurons which terminate in the medulla— these are termed the corticobulbar pathway. This pathway is responsible for controlling the motor neurons that innervate structures in the head and neck (e.g. the tongue).
- Situated above the medulla is the pons. In many respects the role of the pons is broadly similar to that of the medulla.
  - The pons has centres which are concerned with the regulation of ventilation.
  - It is the site of origin of a further four cranial nerves.
- The uppermost portion of the brainstem is the midbrain.
  - The midbrain is the site of origin of three cranial nerves.
  - It contains a number of nuclei concerned with motor activity (e.g. the substantia nigra).
  - The midbrain contains two paired regions known as the inferior and superior colliculi. The inferior colliculi receive an input from auditory neurons and are responsible for the startle reflexes, which respond to sound. The superior colliculi receive an input from the visual system and are responsible for eye movements (e.g. fixation of gaze).
  - Scattered throughout the brainstem is the **reticular formation**—this represents a functional region of the brain rather than a simple anatomical region.

- Neurons are found throughout the brainstem, and indeed in other regions of the brain which contribute to this structure.
- The reticular formation plays a role in a number of aspects of brain function, including those determining the level of arousal and the sleep–wake cycle and also a number of reflexes (e.g. baroreceptor reflexes).

CNS—the diencephalon

- The diencephalon is formed from the **thalamus** and the **hypothalamus**.
- The thalamus plays an important role in the processing of all sensory information except olfaction. It also participates in neural activity related to movement and emotional behaviours.
- The hypothalamus lies beneath the thalamus and is involved in a number of important physiological responses including:
  - thermoregulation
  - the regulation of eating and drinking
  - an interface between the nervous and endocrine systems—a number of hormones are produced here which influence endocrine activity in the pituitary gland.

CNS—the cerebellum

- The **cerebellum**, which is Latin for 'little brain', is located posterior to the brainstem and inferior to the cerebrum.
- Despite its relatively small size, this region of the brain is densely packed with neurons—50% of the neurons within the brain are found here.
- The cerebellum has a number of relatively sophisticated functions.
  - It acts as a comparator, ensuring that movements directed by the motor cortex actually occur. It does this by comparing feedback from proprioceptors with the plan of movement initiated by the motor cortex.
  - It is involved in the learning of new motor skills.

CNS—the cerebrum

- The **cerebrum** is the largest part of the brain. It is divided into two hemispheres (connected at the corpus callosum), which are each divided into four lobes: parietal, occipital, frontal, and temporal.
- The outermost layer (2–4mm thick) of the cerebrum is termed the **cerebral cortex**—this is the grey matter—on account of the neurons here being non-myelinated. The cortex is characterized by a highly folded appearance—the 'peaks' of the folds are known as gyri, whilst the 'troughs' are known as sulci.
- Cortical regions undertake the highest aspects of neural activity within the nervous system. They are involved in the perception of sensory information, in planning voluntary movement, in language, and in sophisticated cognitive functions.

- Each aspect of neural activity is undertaken by a specific region of the cortex—there are cortical regions that deal with visual, auditory, olfactory, and somatosensory information, and so on. There are also regions devoted to the planning and initiation of voluntary movement. Associated with each of these regions are areas of association cortex, which have an integrative region, i.e. combining neural activity from several sources to ultimately generate an overall neural response.

- Lying beneath the cortex in the white matter of the cerebrum are subcortical nuclei. These are extensively interconnected and also interconnect with regions in the thalamus and the brainstem. Together they form the basal ganglia whose role is in the execution of voluntary movements. This arrangement is also known as the extrapyramidal motor system.

CNS—the spinal cord

- The **spinal cord** originates from the medulla of the brainstem. In humans it is about 45cm long and about 1.5cm in diameter. Thus the cord does not extend through the complete length of the vertebral column which encloses it.

- Originating from the spinal cord are 31 pairs of spinal nerves—these branch extensively and ultimately form the peripheral nervous system. The overall organization of the spinal cord can be seen in Figure 2.9. The spinal nerves are named after the region of the vertebral column from which they arise. Thus there are cervical thoracic, lumbar, sacral, and coccygeal spinal nerves.

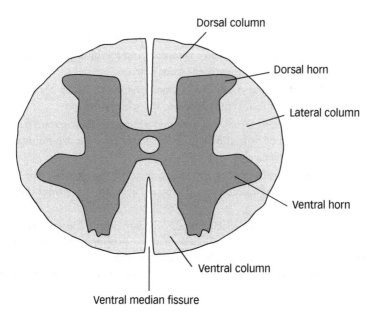

Figure 2.9 Cross-section of the spinal cord

Organization of the nervous system

- In cross-section, the spinal cord is roughly circular. It consists of a central butterfly shaped grey region surrounded by white matter.
- At the level of the spinal cord, there is separation of sensory and motor neurons. Sensory neurons with an input to the spinal cord arise from a range of sensory receptors (e.g. pain, temperature, touch, etc.). Their **dendrites** are in the periphery of the body. Their cell bodies are collected together in the **dorsal root ganglia**, whilst their axons terminate in the grey matter of the spinal cord.
- The grey matter can be divided into a number of layers, each with a specific function. For example, layers 1–6 in the dorsal horn are where the axons of the sensory neurons terminate.
- Motor output from the spinal cord (e.g. to skeletal muscle) leaves via the ventral roots. The dorsal and ventral roots merge to form spinal nerves which, since they contain both sensory and motor neurons, are said to be mixed nerves.
- The grey matter of the spinal cord acts as an 'integrating centre' for **spinal reflexes**. Reflexes are relatively simple automatic predefined responses to stimuli.
- The basic organization of a reflex is as follows: receptor → afferent neuron → spinal cord → efferent neuron → effector.
- Typical spinal reflexes include tendon or myotatic reflexes such as the knee jerk reflex.
  - In this reflex tapping of the patella tendon results in stretching of the quadriceps muscles of the thigh. This stretch is detected by specialized muscle receptors called muscle spindles.
  - Sensory neurons associated with these receptors terminate in the spinal cord and make direct synaptic contact with motor neurons which innervate the quadriceps muscles, causing them to contract and inhibiting the antagonistic hamstring muscle—consequently the knee jerks.
  - The knee jerk reflex is an example of a monosynaptic reflex, i.e. there is a single synapse connecting the afferent neuron to the sensory neuron.
- The white matter contains bundles of axons which convey information from the periphery to the brain or from the brain to the periphery—the ascending and descending tracts, respectively.
- The tracts are named from their origin and destination. For example, the spinocerebellar tract is an ascending tract conveying information from the spinal cord to the cerebellum.
- The tracts themselves convey specific types of information. The spinocerebellar tract conveys information to the cerebellum about proprioceptive input, i.e. information about the position of joints, tendons, etc. The spinothalamic pathway provides information to the thalamus about pain, temperature, and touch.

The spinal cord is shorter than the vertebral column. The nerves from the lower region of the cord form a structure called the cauda equina. This gives the appearance of paired spinal nerves originating from all regions of the vertebral column, even in the absence of a spinal cord.

Although reflexes are automatic responses, in certain circumstances they may be overridden by higher regions of the brain.

Reflexes can be far more complex than the simple stretch reflex described. They may involve a number of neurons, i.e. be polysynaptic, and involve both sides of the body. A good example of this is the crossed extensor reflex.

2.4 SENSORY INPUT

The CNS receives sensory information from both the internal and external environment. It is possible to distinguish two types of sensory input—**somatosensory** and **special senses**. Somatosensory input arises from within the body and signals such information as temperature, touch, pressure, pain, and proprioception (i.e. information about body position). Special senses are focused in the head and include vision, taste, smell, hearing, and vestibular sensation. This section will focus on the general principles of sensory receptor activation, rather than a consideration of any specific sensory structure.

Transduction of sensory information

- The role of sensory receptors is to transduce sensory information—i.e. to convert one source of energy to another. Therefore photoreceptors in the eye 'convert' light to action potentials and hair cells in the ear 'convert' sound waves to action potentials.
- Sensory receptors can be classified in several ways.
  - Interoreceptors gather information about changes in the internal environment (e.g. pressure of blood in the arterial system).
  - Exteroreceptors gather information about the external environment (e.g. photoreceptors in the retina).
  - Receptors can be classified by the type of stimulus they detect.
  - **Mechanoreceptors** detect changes in mechanical deformation. This may include touch and pressure in the skin, which are detected by structures called Meissner's corpuscle and Merkel's disc, respectively. Mechanoreceptors also include the baroreceptors found in the walls of some arteries and which monitor blood pressure.
  - **Chemoreceptors** detect changes in the composition of the body. They include chemoreceptors which detect changes in O_2 and CO_2 and signal appropriate changes to ventilation.
  - **Thermoreceptors** detect changes in temperature. These receptors are capable of detecting both cold and hot.

- ○ Some receptors can be considered to be polymodal, i.e. they respond to more than one type of stimulus. **Nociceptors**, which are receptors responsible for detecting pain, are a good example of this as they respond to both mechanical and chemical stimuli.
- The process of **transduction** involves the production of a receptor potential. Receptor potentials are localized potentials, i.e. they are transient, they decay as they move away from the point of initiation, and they are able to summate. In some receptors, the receptor potential will initiate the production of full APs, which are conveyed back to the CNS. In other receptors, it is the receptor potential itself which is the signal of change.
- In the face of prolonged stimulation, sensory receptors display a reduction in their response, i.e. they are said to adapt.
  - ○ **Rapidly adapting receptors** show a rapid reduction in their output—thermoreceptors display this type of activity.
  - ○ **Slowly adapting receptors** show a much slower reduction in their activity in response to prolonged stimulation—nociceptors are a good example of this phenomenon.
- Each receptor responds preferentially to a given stimulus, which is sometimes called the adequate stimulus.
- In responding to stimulation, it is both the number of receptors which are activated and the frequency of action potentials generated which ultimately give rise to the perception and magnitude of the stimulus.
- Somatosensory information enters the spinal cord via the dorsal roots of spinal nerves and is transmitted to the brain via the spinothalamic tracts. Primary afferent fibres travel in the dorsal columns of the white matter. They synapse in the medulla with second-order neurons, which decussate and travel to the thalamus. Here they synapse with third-order neurons, which terminate in the somatosensory cortex of the cerebrum.

2.5 MOTOR OUTPUT FROM THE CNS

Motor output from the CNS brings about the activation of effectors (e.g. activation of skeletal muscle), producing an overt response. Alternatively, the **autonomic nervous system (ANS)**, which initiates change internally (e.g. contraction of gut smooth muscle to aid in the process of digestion), may be activated.

Somatic motor output

- Somatic motor output refers to the activation of skeletal (voluntary) muscle.
- Production of motor activities is a complex process involving several regions of the brain. Much human movement is initiated voluntarily, but some aspects occur relatively automatically once initiated (e.g. walking).

- The planning of motor activity begins in the premotor cortex. This is the cortex associated with the motor cortex.
- This plan, or intention to move, is relayed to the motor cortex itself. By itself, the plan to initiate movement is not enough. Two other brain regions are important—the basal ganglia and the cerebellum.
- The role of the cerebellum is to integrate the timing and coordination of the desired movements.
- The **basal ganglia** are required to activate the appropriate motor cortical neurons necessary for movement, whilst at the same time inhibiting unwanted motor cortical activity.
- The cerebellum and basal ganglia feed back to the motor cortex via the thalamus.
- Finally, the necessary output of the motor cortex is relayed to nuclei in the brainstem which give rise to descending motor pathways—the medial pathways and the lateral pathways.
- The medial pathways are concerned with the maintenance of an appropriate posture, amongst other things.
- The lateral pathways are pathways from the motor cortex to the motor neurons in the spinal cord, which, when activated, produce movement in the head and limbs.

Autonomic output

- Whilst activation of the somatic motor system produces overt movement, activation of the autonomic nervous system (ANS) produces movement which is not recognized at the level of consciousness and which for the most part is not seen.
- The ANS is effectively the motor system which produces responses in the viscera—essentially, smooth muscle responses—in order to ensure visceral homeostasis.
- The ANS is divided into two divisions—the **sympathetic ANS (SANS)** and the **parasympathetic ANS (PANS)**. Generally, these two divisions have opposing actions—SANS is considered to be the 'fear, flight, fight' system, while PANS is considered to be the 'rest and digest' system.
- Unlike the somatic motor pathway described previously, the pathway between the CNS and effectors in the ANS consists of a two-neuron chain (Figure 2.10).
  - The pre-ganglionic neurons have their cell bodies in the CNS. Their axons leave via the ventral roots and travel in the spinal nerves. They leave the spinal cord and synapse with post-ganglionic neurons. Some pre-ganglionic neurons travel in cranial nerves rather than spinal nerves.
  - Post-ganglionic neurons have their cell bodies located in peripheral autonomic ganglia, and their axons terminate on effector structures—smooth muscle of the viscera and blood vessels.

Motor output from the CNS

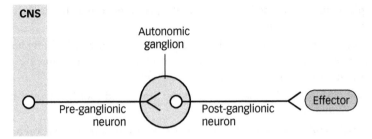

Figure 2.10 The arrangement of neurons in the autonomic nervous system

- Pre-ganglionic neurons of the SANS leave the spinal cord via thoracic and lumbar spinal nerves. They release the neurotransmitter acetylcholine, which stimulates the post-ganglionic neurons. The post-ganglionic neurons release the neurotransmitter noradrenaline onto effector structures.
- Pre-ganglionic neurons of the PANS leave the spinal cord via the sacral spinal nerves and also some of the cranial nerves. They release the neurotransmitter acetylcholine onto the post-ganglionic neurons. Post-ganglionic neurons release the neurotransmitter acetylcholine onto effector structures.
- Internal structures may have varying patterns of innervation from the SANS and the PANS.
  - Dual antagonistic innervation (e.g. the heart)—activation of the SANS produces an increase in heart rate and force of contraction, while activation of the PANS produces the opposite response.
  - Single innervation (e.g. vascular smooth muscle in arteries and arterioles)— here, there is a single innervation from the SANS, which, when activated, produces vasoconstriction.
  - Dual non-antagonistic innervation (e.g. the male reproductive system)—in this case-activation of the PANS is involved in the erectile response of the penis whilst activation of the SANS is necessary for ejaculation.

Check your understanding

Distinguish between the ionic mechanisms responsible for the resting membrane potential and action potentials. *(Hint: think about the ionic composition of ICF and ECF)*

Why does 50ml of room temperature water in a beaker feel lighter than 50ml of cold water when placed on the forehead? *(Hint: consider the idea of the adequate stimulus and cross-activation of receptors)*

Describe the basic organization of a reflex pathway. *(Hint: review the section 'Spinal cord' in section 2.3)*

3 The endocrine system

Together with the nervous system, the endocrine system controls the multitude of physiological processes in the body, which ultimately ensure that homeostasis is achieved. There are significant differences in the systems, but also a number of instances where the division between them is becoming increasingly blurred. So, although it may be convenient to consider them as two separate systems, in reality life is far more complex.

Key concepts

- Together with the nervous system, the endocrine system works to control and coordinate a tremendous variety of physiological processes.
- Hormones may be formed from derivatives of amino acids, peptides and proteins, and lipids. The chemical nature of an individual hormone determines where its receptor is located—membrane or intracellular.
- Although hormone secretion may display **circadian rhythms**, overall plasma levels are maintained at the appropriate level by negative feedback mechanisms.
- There are a number of organs with endocrine activity—some are more established and better understood than others.
- The pituitary gland is sometimes described as the master endocrine gland by virtue of the number of physiological processes it controls.
- Pathophysiology, i.e. over- or underactivity of endocrine glands, provides an insight into normal physiological function.

3.1 CHEMICAL VERSUS NEURAL CONTROL

Both the endocrine system and the nervous system provide control and coordination of physiological systems, but there are important differences between them.

- Endocrine control provides a chemically mediated control system in the body alongside the nervous system.
- Control exerted by the endocrine system differs in a number of respects from that exerted by the nervous system:
  - nature of the control—endocrine exerts its control by releasing chemicals whereas the nervous system uses electrical control;
  - rapidity of the control response—endocrine control responses may last for weeks, months, or years (e.g. growth) in contrast with the very short-lived responses of the nervous system (e.g. reflexes);
  - duration of the controlling signal—endocrine signals may take hours or days to initiate responses (e.g. steroid hormone action) in contrast with the very rapid duration of neural signals (i.e. action potentials).
- Despite the obvious differences between the two control systems, it is important to realize that they work cooperatively to achieve overall control.
- There are some aspects of control (e.g. neuroendocrine control) which share elements of both systems. A typical example of this is control of water balance mediated by the pituitary hormone vasopressin.

3.2 'TYPES' OF ENDOCRINE CONTROL

There are different types of endocrine control—classification is based on where the hormone is released from (neural or non-neural tissue) and whether it enters the general circulation.

- Endocrine control is characterized by the release of chemical substances—hormones—which interact with target cells to produce biological responses.
- Endocrine control can be considered to consist of three categories—**autocrine control**, **paracrine control**, and classical **endocrine control**. These are summarized in Figure 3.1.
- Autocrine control is characterized by the release from a cell of a chemical which interacts with and influences the activity of the cell from which it was released.
- Paracrine control is characterized by the release of a hormone from a cell into the interstitial fluid surrounding the cell. The hormone, remains in the interstitial fluid, and diffuses a short distance where it influences the activity of a target cell.
- Classical endocrine control is the most common form of control. In this case, a hormone is released from a cell, enters the circulatory system, and travels some distance to influence biological activity in a target cell.
  - The cells which release hormones in classical endocrine control are aggregated into endocrine organs.

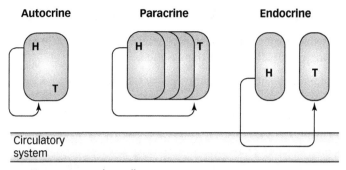

H – Hormone secreting cell
T – Target cell

Figure 3.1 Summary diagram illustrating the 'types' of endocrine control

○ Endocrine organs may be composed of epithelial tissue and form classic endocrine organs (e.g. pancreas, thyroid).
○ Alternatively, endocrine organs may be composed of neural tissue (e.g. posterior pituitary gland). In this case the hormone is released from the axon terminals of neurons. This is termed **neuroendocrine control**.

3.3 THE CHEMICAL NATURE OF HORMONES

Hormones are classified as being either water-soluble (amino acid derivatives, peptides, and proteins) or lipid-soluble (steroids)—this will determine where the receptor, with which the hormone interacts to produce its biological effects, resides.

• The chemical nature of an individual hormone is important because this determines 'how' the hormone exerts its effects. The chemical nature of the hormone will determine whether it can enter the target cell.
• Hormones fall into one of two chemical groups:
  ○ proteins, peptides, and modified amino acids, e.g. **insulin**, **oxytocin**, and **adrenaline**, respectively;
  ○ steroids (e.g. **cortisol**), which are formed from modifications of a cholesterol molecule.
• Proteins, peptides, and modified amino acids are water-soluble compounds. Therefore they are unable to cross the lipid-rich membranes of target cells. Because of this their receptors are located in the membranes of target cells.
• Steroids are lipid-soluble compounds. Therefore they are able to cross the membrane of target cells. Steroid receptors are located in the cytoplasm of target cells.

3.4 MECHANISM OF ACTION OF HORMONES

Hormones produce their effects indirectly by influencing either the production of second-messenger molecules or gene transcription.

- The presence (or absence) of receptors ensures that hormones are selective in their action: no receptor = no activity.
- Proteins, peptides, and modified amino acids bind (reversibly) to receptors located in the membrane of target cells.
- Once bound, the hormone–receptor complex interacts with, and activates, a **G-protein**. G-proteins are proteins found in cell membranes which act as intermediaries between hormone binding and the initiation of responses within the target cell.
- Subsequently, the activated G-protein activates an enzyme called **adenylate cyclase**. This enzyme is bound to the membrane of the target but faces into the cytoplasm.
- Adenylate cyclase takes a molecule of adenosine triphosphate (ATP) and converts it to a molecule of **cyclic AMP** (**cAMP**). cAMP is known as a second-messenger molecule—the first messenger is the hormone itself.
- The rise in cAMP levels converts inactive intracellular **protein kinases** to active protein kinases. Protein kinases are enzymes which phosphorylate other (target) proteins. When a protein is phosphorylated, its conformation (shape) changes.
- The now activated protein kinases phosphorylate other intracellular proteins. These proteins may be, for example, enzymes or membrane-bound transport proteins. Since the conformation of these proteins changes, so does their function. therefore an inactive enzyme may be changed to its active form and the target cell may now start producing a chemical. This is the biological response of the target cell.
- Activation of the target cells needs to be regulated—having been switched on by a hormone, it now needs to be switched off. Activity in the target cell is achieved by the conversion of cAMP to AMP. This is done by the enzyme **phosphodiesterase**.
- The processes described here produce responses very quickly. Consider, for example, the immediate effects seen when adrenaline is released into the blood—an increased heart rate and dilation of the pupils, amongst other responses, which occur almost instantaneously.
- A summary of the action of hormones which have receptors located in the cell membrane can be seen in Figure 3.2.
- The action of **steroid hormones**, which have intracellular receptors located in the cytoplasm, is a little simpler.
- Steroids are lipid-soluble—therefore they do not easily dissolve in aqueous solutions (e.g. plasma in the body). This means there is a potential difficulty in transporting them around the body from their site of production to their site of action.

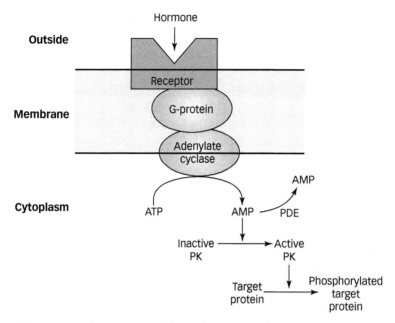

Figure 3.2 The action of hormones with membrane-bound receptors

- In order to overcome this problem, steroids are transported in blood plasma bound to transport proteins, known as **steroid-binding proteins**, which are found in the plasma.
- At the target cell, the steroid dissociates from the transport protein and, as it is lipid-soluble, passes through the cell membrane and enters the cytoplasm of the target cell.
- In the cytoplasm of the target cell it binds to its receptor.
- The hormone–receptor complex then passes through the nuclear membrane of the target cell where it binds to specific regions of DNA.
- The outcome of binding to DNA is that genes are activated—either on or switched off. This results in the production of proteins (e.g. enzymes), which alter the biological activity of the target cell. This represents the response of the target cell to the hormone.
- Within the target cell, there are several mechanisms which limit the activity of steroids. These include a finite number of receptors to which the steroid can bind, and in some cases the steroid may actually inhibit the synthesis of new receptors.
- Compared with hormones, which have membrane-bound receptors, the response of cells to steroids is much slower (hours to days to produce effects) as it involves the activation of genes, production of proteins, post-translational modification of proteins, etc.
- A summary of steroid hormone action can be seen in Figure 3.3.

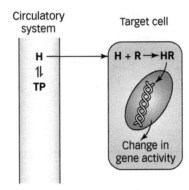

Figure 3.3 The action of hormones with intracellular receptors

3.5 REGULATION OF HORMONE LEVELS

Plasma levels of hormones need to be closely regulated to ensure that homeostasis is maintained. Too little or too much of a hormone is associated with pathophysiology.

- It is essential that appropriate levels of hormones are maintained within the plasma—too little or too much of an individual hormone will have deleterious effects on normal physiological function. For example, diabetes is a consequence of too little of the pancreatic hormone insulin.
- Overall levels of an individual hormone represent the balance between its secretion and degradation.
  - Secretion of hormones occurs in response to a stimulus. For example, the appearance of glucose in plasma following a meal results in the production of insulin.
  - Superimposed upon this in the case of some hormones are **circadian changes** in secretion. For example, the release of growth hormone is at a maximum during the early stages of sleep.
  - Degradation of hormones may occur by several mechanisms: metabolism in the liver, renal loss, and catabolic destruction in tissues. This may result in some hormones having extremely short half-lives—seconds in the case of some of the prostaglandins. On the other hand, since steroids are bound to proteins they tend to have much longer half-lives (days).
- For the most part, levels of individual hormones are controlled by negative-feedback mechanisms—a rise in the plasma level of a hormone inhibits further secretion of it, thus allowing the plasma level to return to its normal level.

3.6 THE PITUITARY GLAND AND HYPOTHALAMUS

The hypothalamo-pituitary axis is a dominant control system in endocrine control, controlling a significant number of other endocrine organs.

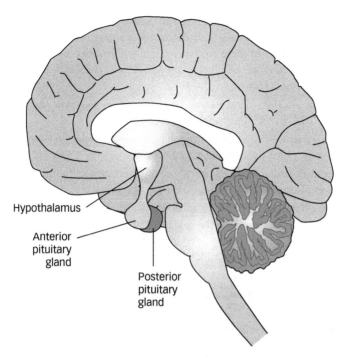

Figure 3.4 The relationship between the hypothalamus and the pituitary gland

- The close link between endocrine and neural regulation, indicated earlier, is best demonstrated when considering the relationship between the hypothalamus and the **pituitary gland**.
- The pituitary gland oversees the activity of a number of endocrine glands. It is sometimes referred to as the 'conductor of the endocrine orchestra'.
- The pituitary gland is located at the base of the brain in a bony structure called the sella turcica. It is connected to the hypothalamus by the **hypophyseal stalk**. This relationship is shown in Figure 3.4.
- The pituitary gland is divided into two distinct regions—the **posterior pituitary gland** and the **anterior pituitary gland** (Figure 3.4). These two regions have distinct origins—the posterior pituitary gland is composed of neural tissue (formed from an outgrowth of the hypothalamus) whereas the anterior pituitary gland is formed from pharyngeal epithelial tissue. This produces clear distinctions in how the regions function.

The posterior pituitary gland

- The posterior pituitary gland represents the endings of neurons which have their cell bodies in the hypothalamus (Figure 3.5).

The pituitary gland and hypothalamus

- Stimulation of these neurons results in the release of substances from these endings directly into plasma—this is known as neurosecretion. In this case, the neurotransmitters in these axons are actually hormones. Unlike normal neurotransmission, rather than being released into a synapse, these substances are released into plasma (Figure 3.5).
- The posterior pituitary gland secretes two hormones—**oxytocin** and **vasopressin**. Both substances are small peptide molecules.
- Oxytocin is released from neurons that have their cell bodies in a region of the hypothalamus called the **supra-optic nucleus.**
- Oxytocin stimulates milk ejection from the mammary glands and also causes contraction of uterine smooth muscle during childbirth.
- In the case of milk ejection, stimulation of the nipple by a feeding child generates action potentials that travel via the spinal cord to the hypothalamus. Here they activate neurons in the supra-optic region, which results in secretion of oxytocin into the blood and ultimately milk ejection.

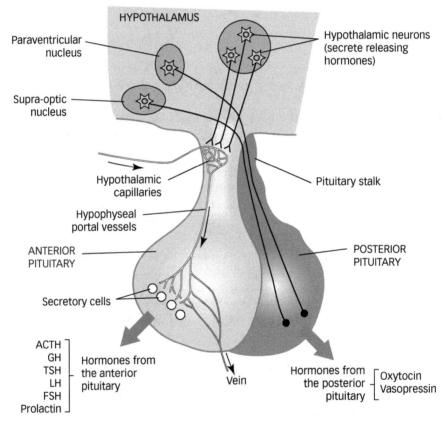

Figure 3.5 Projection of neurons from the hypothalamus to the posterior pituitary gland

- Vasopressin is secreted from neurons whose cell bodies lie in the **paraventricular nucleus** of the hypothalamus. The principal role of vasopressin is regulation of water balance.
- The primary stimulus for its release is an increase in plasma osmolarity. This change is detected by osmoreceptors in the hypothalamus which in turn activate neurons, which release vasopressin.
- Vasopressin acts primarily in the collecting duct of the nephron to increase water reabsorption and consequently the osmolarity of the plasma is reduced.

The anterior pituitary gland

- In contrast to the neurosecretion seen in the posterior pituitary gland, the anterior posterior pituitary gland can be considered to be a more typical endocrine structure, i.e. non-neural tissue secreting hormones into the plasma.
- The hypothalamus controls the secretions of the anterior pituitary gland itself by the release of hormones—so-called **release hormones (RH)** and **release-inhibiting hormones (RIH)**. These hormones are released by cells in the hypothalamus into the plasma where they travel to the anterior pituitary gland to exert their effects.
- The blood vessels which connect the hypothalamus to the anterior pituitary gland are known as the **portal blood vessels**. These are vessels which having passed through one organ pass through another before returning to the heart.
- The release and release-inhibiting hormones leave the circulation and enter the anterior pituitary gland where they stimulate or inhibit the release of anterior pituitary hormones. This arrangement can be seen in Figure 3.6.

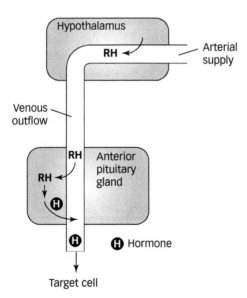

Figure 3.6 The role of portal blood vessels in transporting releasing factors to the anterior pituitary gland

- The release and release-inhibiting factors secreted from the hypothalamus are:
  - **luteinizing hormone releasing hormone (LHRH)**
  - **thyrotrophin releasing hormone (TRH)**
  - **growth hormone releasing hormone (GHRH)**
  - **corticotrophin releasing hormone (CRH)**
  - **prolactin inhibiting factor (PIF)** (also known as dopamine)
  - **growth hormone inhibiting hormone (GHIH)** (also known as somatostatin).
- In turn, these hypothalamic hormones stimulate or inhibit the release of the following anterior pituitary hormones:
  - LHRH—stimulation of **follicle stimulating hormone (FSH)** and **luteinizing hormone (LH)** release
  - TRH—stimulation of **thyroid stimulating hormone (TSH)** release
  - GHRH—stimulation of **growth hormone (GH)** release
  - CRH—stimulation of **adrenocorticotrophic hormone (ACTH)** release
  - PIF—inhibition of prolactin release
  - GHIH—inhibition of GH release.
- Each of these anterior pituitary hormones then influences activity in other endocrine organs:
  - FSH and LH are collectively termed gonadotrophins. They are involved in the regulation of reproduction and their actions will be described in Chapter 9.
  - TSH, also known as thyrotrophin, stimulates the thyroid gland to release the hormones T_3 and T_4.
  - GH influences both metabolism (e.g. it promotes fatty acid oxidation) and growth in a more overt sense (e.g. it promotes cartilage formation).
  - ACTH regulates the function of the adrenal glands—the small glands lying at the superior pole of each kidney.
  - PIF inhibits the release of prolactin.
- It can be seen that there is a close working relationship between the hypothalamus and anterior pituitary gland—this is sometimes referred to as the hypothalamo-pituitary axis.

3.7 THYROID HORMONES

Thyroid hormones released from the thyroid gland influence a number of physiological processes, including control of metabolic rate.

- The **thyroid gland** is a bi-lobed gland located in the neck, just below the larynx.
- It is a relatively small gland, but has a dense blood supply.

- The functional unit of the gland is the thyroid follicle—the complete gland is made up of several thousand follicles. The follicle is characterized by its ability to absorb iodine from dietary intake in the form of the iodide ion.
- The follicle concentrates iodide ions and oxidizes them to convert them to iodine. The iodine is then linked to tyrosine amino acid residues, which are themselves linked to a protein called thyroglobulin which is found within the cytoplasm of the follicle—this forms mono-iodotyrosine (MIT). Subsequently, a second iodine molecule is attached– this forms di-iodotyrosine (DIT).
- Coupling between MIT and DIT results in the formation of the active substances **tri-iodothyronine (T_3)** and **tetra-iodothyronine (T_4)**.
- Under the influence of TSH thyroglobulin is hydrolysed, releasing T_3 and T_4 into the circulatory system. The hormones are transported around the body bound to plasma proteins. Both hormones enter target cells by active transport and, once inside the cell, T_4 is converted to T_3.
- T_3 and T_4 influence virtually all cells in the body. They are involved in the following processes.
  - Production of heat—this is an important factor in controlling body temperature. In terms of whole-body responses, this is reflected by an increase in basal metabolic rate.
  - Low levels of the hormones appear to stimulate the absorption of amino acids and protein anabolism, whilst high levels have the opposite effect.
  - They promote the breakdown of fat stores, resulting in a rise in the plasma concentrations of free fatty acids.
  - Likewise, low levels stimulate the uptake of glucose into muscle cells, with an insulin-like effect. High levels of the hormones promote the breakdown of glucose stores and the synthesis of new glucose.

Looking for extra marks?

Although T_3 and T_4 are amino-acid-based hormones, they are unusual in that they have intracellular receptors rather than receptors in the membrane of target cells. It is now thought that there may be receptors for them in mitochondria—this may help explain their influence on metabolic rate.

Hyperthyroidism is an overactivity in the thyroid gland. The thyroid swells (goitre) and the sufferer displays a loss of weight, excessive sweating, and arrhythmia, amongst other symptoms. It can be treated by surgical removal of parts of the thyroid or by the administration of radioactive iodine, which is incorporated into the gland and destroys it.

Hypothyroidism is the result of an underactive thyroid gland. Amongst the symptoms, sufferers have a swollen face, are lethargic, and show a gain in weight. It is treated by the administration of either iodine supplements or thyroid hormones. In infants, severe hypothyroidism can lead to severe mental impairment called cretinism. This is because thyroid hormones are essential for normal brain development.

3.8 PARATHYROID HORMONE

The parathyroid glands are important regulators of plasma calcium levels. This is essential given the key role that calcium has as a signalling substance within the body.

- Whilst approximately 99% of the total body calcium is contained within the skeletal system, there is an extensive control system which regulates the plasma concentration of the remainder. This is because of the role that calcium plays in a variety of physiological processes (e.g. muscle contraction and secretory processes).
- The **parathyroid glands** are embedded in the thyroid gland. They release a hormone called **parathyroid hormone** (**PTH**), which regulates, in part, plasma calcium levels. PTH increases plasma calcium levels and decreases plasma phosphate levels. Its targets are the skeleton, gastrointestinal tract, and kidneys.
- In healthy individuals with a normal plasma calcium concentration, PTH promotes activity in **osteoblasts** (bone-building cells) and the calcification of bone.
- When plasma calcium levels drop, PTH stimulates the release of calcium, initially from the surface of the bone, but in the long term it also activates **osteoclasts** (bone-destroying cells). This results in a rise in plasma calcium concentration level.
- In the gastrointestinal tract, PTH indirectly stimulates the absorption of calcium. It does this by promoting the production of a compound called 1,25-dihydroxy-cholecalciferol. This is synthesized from vitamin D and promotes the intestinal absorption of calcium.
- In the kidney, PTH stimulates calcium reabsorption in the distal convoluted tubule. At the same time, it decreases the reabsorption of phosphate. The significance of the reduction in plasma phosphate levels is that calcium binds to phosphate. Therefore reducing phosphate concentrations results in a further rise in calcium levels.
- PTH secretion is controlled by plasma levels of calcium. Low plasma calcium levels promote its release, whilst high levels inhibit its release.

Looking for extra marks?

In addition to PTH and 1,25 dihydoxy-cholecalciferol, Ca^{2+} ions are also regulated by another hormone called **calcitonin,** which is released from the thyroid gland. Its primary role is the inhibition of osteoclasts—in doing this, it reduces plasma calcium levels.

3.9 GROWTH HORMONE

As the name indicates, growth hormone is responsible for the control of growth. Whilst it is active throughout all life stages, there are periods when its role is increased (e.g. childhood and adolescence).

- Growth hormone is released from the anterior pituitary gland and influences virtually all cells in the body.

- Despite its major role as a stimulator of growth in young children and adolescents, it retains an influence throughout adult life. The actions of growth hormone can be divided into direct and indirect actions.
  - ○ The primary direct action of growth hormone is its stimulatory effect on protein synthesis. These effects are of particular significance during childhood growth—especially in bone. It also stimulates the growth of cartilage. In promoting growth it also produces a glucose-sparing effect—promoting ATP production from fatty acid oxidation whilst conserving glucose.
  - ○ The primary indirect action of growth hormone is the maintenance of tissue.
  - ○ The indirect effects of growth hormone are, in part, mediated via the production of substances called **insulin-like growth factors** (**IGFs**) in the liver. The principal IGFs are IGF-1 and IGF-2.
  - ○ IGF-1 plays a significant role in growth, and amongst other roles stimulates mitotic cell division, whilst IGF-2 has an insulin-like activity.
- The release of growth hormone from the anterior pituitary gland is influenced by two hormones from the hypothalamus—growth hormone releasing hormone and somatostatin—GHRH is the most important factor. The release of GHRH is influenced by a number of stimuli to the hypothalamus (e.g., pain, exercise, and reduced blood sugar levels).

Looking for extra marks?

GH, like other anterior pituitary hormones, displays a characteristic pattern of release. It is secreted in a pulsatile manner. Activity is greatest during periods of significant growth (e.g. during adolescence). Against this pulsatile nature, there is a characteristic diurnal pattern of secretion, with increased secretion occurring during sleep.

Overproduction of GH in childhood results in gigantism, whilst underproduction results in pituitary dwarfism. In adults, overproduction of GH results in acromegaly—a condition where the extremities (e.g. the hands) are enlarged.

3.10 THE ADRENAL GLANDS

The **adrenal glands** are multifunctional glands which are involved in a variety of physiological processes (e.g. metabolism, and fluid and electrolyte balance).

- The adrenal glands are a pair of glands—one lying above each of the kidneys. Each gland consists of two regions—the adrenal medulla and surrounding this the adrenal cortex (Figure 3.7).

The adrenal medulla

- The adrenal medulla is the inner part of the adrenal gland. It is effectively a specialized ganglion of the sympathetic nervous system. The principal cells of the

The adrenal glands

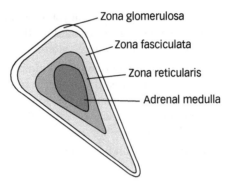

Figure 3.7 Cross-section of the adrenal gland

adrenal medulla are the chromaffin cells. When the sympathetic nervous system is stimulated, they release two compounds—adrenaline and **noradrenaline**.

- These two compounds have common cardiovascular effects (e.g. increased heart rate and force of contraction). Adrenaline also has more selective actions (e.g. vasodilation in skeletal muscle vascular beds and bronchoconstriction).
- Adrenaline has significant effects on metabolism (e.g. it stimulates hepatic glycogenolysis).
- Sympathetic activity, and therefore activity in the gland, is increased during a variety of states (e.g. exercise and pain).

The adrenal cortex

- The adrenal cortex surrounds the medulla and is itself divided into three distinct regions (Figure 3.7). Together, these regions secrete a variety of hormones, all of which are derivatives of cholesterol, i.e. they are steroids.
- The outermost region, the **zona glomerulosa**, secretes mineralocorticoids.
  - The principal mineralocorticoid is the hormone **aldosterone**.
  - The secretion of aldosterone is regulated by the renin–angiotensin system (see Chapter 7).
  - A decrease in the plasma level of sodium activates this system. In response, aldosterone is released, which acts to increase the reabsorption of sodium in the kidney, thus restoring the correct plasma concentration.
- The middle region, the **zona fasiculata**, secretes glucocorticoid hormones.
  - The principal glucocorticoid is cortisol—others include corticosterone and cortisone.
  - Together, these hormones contribute to the ability of the body to cope with a variety of stressors (e.g. trauma and infection). They display anti-inflammatory and immunosuppressive actions amongst others.
  - The other significant role for cortisol is in the regulation of carbohydrate and protein metabolism—in essence; actions here tend to be the opposite of the

actions of insulin. One of its main actions is to promote the conversion of protein to glycogen

○ Glucocorticoids are controlled by the anterior pituitary hormone ACTH which, in turn, is controlled by the hypothalamic hormone CRH. As with all hormonal systems, the process of negative feedback regulates its secretion.

• The inner region, the **zona reticularis** secretes sex hormones.

○ Androgens are produced in males, whilst oestrogens and progesterone are produced in females. Compared with gonadal production of these substances, adrenal production is minor.

○ The precise role of adrenal sex hormones in normal physiology is unclear. However, they may play a role in the development of some of the secondary sex changes occurring during puberty in girls.

Looking for extra marks?

ACTH, which regulates adrenal cortex function, has a characteristic diurnal rate of secretion. For example, plasma cortisol levels peak in the early morning between 6 and 8 a.m. They then fall for the remainder of the day, reaching a minimum in the early hours of the following morning—about 3 a.m. Within this circadian release, there are periods of pulsatile release which cause transient rises in plasma levels.

A number of disease states are associated with over- or underactivity of the adrenal cortex. Overproduction of cortisol is known as Cushing's syndrome. This may be due to increased release of cortisol itself or increased release of ACTH or CRH—all will produce the same effect. Addison's disease is the result of underactivity in the adrenal cortex. This is a rare disease that presents with overall reduction in all hormones of the adrenal cortex.

3.11 PANCREATIC HORMONES

The pancreatic hormones are primarily responsible for regulating plasma glucose levels. Inability to do this results in diabetes mellitus.

• The pancreas secretes two hormones—insulin and **glucagon**—which are important in the maintenance of an appropriate plasma glucose concentration.

• These hormones are secreted from endocrine regions of the pancreas known as the **islets of Langerhans**. A number of cell types (α, β, and δ) have been identified in the islets.

Insulin

• Insulin is produced and stored in the β cells of the islets of Langerhans.

• The principal stimulus to insulin release is an increase in plasma glucose concentration. As the concentration rises, glucose depolarizes the β cells and promotes a calcium-mediated exocytosis of the hormone.

Pancreatic hormones

- In response to an increased plasma glucose concentration (e.g. after consumption of food) there is a biphasic release of insulin. The initial response corresponds to the release of stored preformed insulin, whilst the later response corresponds to the release of newly synthesized hormone.
- The main sites of action of insulin are the liver, skeletal muscle, and adipose tissue. In all tissues, the response is increased active uptake of glucose from plasma into these structures. Having entered, the majority of the glucose is converted to glycogen.
- A variety of other stimuli also control insulin secretion. These include activation of the sympathetic nervous system, which results in a decrease in insulin release. Activity in the parasympathetic nervous system has the opposite effect.

Glucagon

- The actions of glucagon effectively oppose those of insulin, i.e. its release increases the plasma concentration of glucose. Therefore its release is stimulated when plasma glucose levels drop.
- The principal target for its action is the liver where it promotes the breakdown of glycogen, whilst at the same time inhibiting the production of new glycogen.
- Glucagon also promotes the synthesis of glucose from amino acids and the release of free fatty acids from fat sources. The significance of the latter response is that this ensures that the brain continues to receive a supply of glucose, whilst other tissues are able to utilize lipid sources for their energy requirements.

Looking for extra marks?

The inability to regulate plasma glucose levels appropriately results in diabetes mellitus. This may occur early on in life owing to a fundamental defect in the pancreas—type 1 diabetes. This is treated by the administration of insulin. Alternatively, it may occur in latter life—type 2 diabetes. This is treated by a mixture of drugs and changes in lifestyle. Whilst insulin and glucagon are the principal hormones regulating plasma glucose levels, many others (e.g. growth hormone, adrenaline, and glucocorticoid hormones) also exert important influences.

Check your understanding

Describe how the chemical nature of hormones influences the location of their receptors in target cells. (*Hint: consider the solubility of hormones in plasma and membranes*)

Why do steroid hormones have a longer time to action than non-steroidal hormones? (*Hint: think about the mechanism of action of steroids*)

What are the potential sites of dysfunction in an individual with diabetes insipidus? (*Hint: think about the release and action of vasopressin*)

4 Muscle

There are three types of muscle in the human body: skeletal, smooth, and cardiac muscle. These different types of muscle have some overlapping qualities; however, each has distinct anatomical and contractile properties. This chapter aims to provide an overview of these different properties.

Key concepts

- The three types of muscle in the human body differ in some aspects of anatomy and regulation of contraction.
- The interaction of actin and myosin filaments is central to the contraction of all types of muscle.
- Skeletal and cardiac muscle are termed 'striated muscle' because of the existence of sarcomeres.
- Regulation of contraction of all three muscle types depends on the movement of calcium into the cytosol of the cell.
- In skeletal muscle, increases in cytosolic calcium are due to central nervous system activation.
- In smooth muscle, increases in cytosolic calcium can be due to central nervous system activation, but can also be due to local factors, hormones, and spontaneous electrical activation.

continued

- In cardiac muscle, increases in cytosolic calcium are due to the action of the conduction system of the heart.
- The increase in cytosolic calcium results in the interaction of actin and myosin filaments, which leads to muscle contraction.

4.1 SKELETAL MUSCLE

There are approximately 640 skeletal muscles in the human body. They can be hard to classify, which leads to problems in accurately counting the number of these muscles.

- Skeletal muscle is attached to bones via **tendons**. Consequently, when skeletal muscle contracts or relaxes, this leads to movement of the skeleton.
- Skeletal muscles tend to act in pairs. For example, the biceps brachii and triceps brachii work together to flex or extend the elbow joint.
- The smallest skeletal muscle in the body is the stapedius muscle. This is found in the inner ear and stabilizes the stapes bone. It is approximately 1mm in size.
- The longest skeletal muscle in the body is the sartorius muscle, which is found in the thigh. It can be up to 30cm long.
- Movement of the skeleton is the main function of skeletal muscle. However, maintaining posture, support for visceral organs, **thermoregulation**, and storage of glycogen and triglycerides are also important functions.
- Contraction of skeletal muscle can only occur following activation of the central nervous system.

Anatomy

- The anatomy of skeletal muscle can be considered at the organ or cellular level (Figure 4.1).
- At the organ level, the skeletal muscle comprises large numbers of skeletal muscle cells bound together with a layer of connective tissue called the **epimysium**.
- Skeletal muscle has a rich nerve and blood supply.
- Within the skeletal muscle, small bundles of skeletal muscle cells are grouped together within another layer of connective tissue called the **perimysium**. These bundles of cells are called **fascicles**.
- Within each fascicle, there are numerous skeletal muscle cells. Each muscle cell is encased within another layer of connective tissue called the **endomysium**.

Skeletal muscle fibres

- Skeletal muscle cells, or muscle fibres, are very different from other cells. The diameter of a **skeletal muscle fibre** can be up to 100μm.
- The stereotypical cell is round and has a single nucleus that is situated in the middle of the cell.

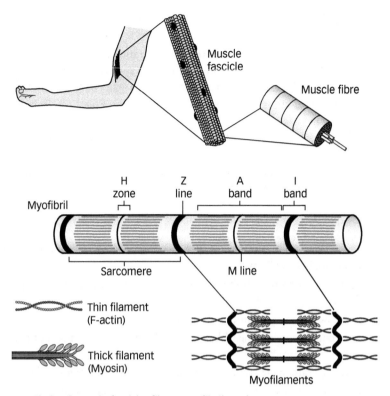

Figure 4.1 Skeletal muscle fascicle, fibre, myofibril, and sarcomere

- Skeletal muscle fibres are cylindrical and can extend the length of the entire skeletal muscle (up to 30cm). In addition, they are multinucleated and these nuclei are found on the periphery of the cell. Skeletal muscle fibres can have a large density of mitochondria, which is beneficial, allowing large degrees of oxidative metabolism to occur within the cell.

- Skeletal muscle fibres also contain a **myosatellite cell**. This ensures that, when damaged, the cell can repair itself. Resistance exercise causes microscopic tears in the skeletal muscle fibre, which are repaired soon after exercise is terminated. The muscle fibre increases in size (**hypertrophies**) due to supercompensation of this tear. Over time, this results in an increase in the cross-sectional area of the organ and a consequent increase in muscle strength.

- Muscle biopsies are often taken by researchers who study skeletal muscle. When a biopsy is taken and examined, a regular pattern of light and dark bands can be seen running along the length of each muscle fibre. This gives skeletal muscle a **striated** appearance. The light and dark bands are due to the arrangement of thin and thick proteins in the cytoplasm of the skeletal muscle fibre. This is called a **myofibril**. A myofibril is approximately 1–2μm in diameter.

- The thin proteins within a myofibril are called **actin** filaments and the thick proteins are called **myosin**.

Skeletal muscle

- The light band is called the I band and the dark band that is seen is called the A band.
- Each myofibril is surrounded by the **sarcoplasmic reticulum**. This is similar to the smooth endoplasmic reticulum found in other cells and is a very rich source of calcium. The sarcoplasmic reticulum is very well developed in skeletal muscle.
- Within each skeletal muscle fibre, **transverse tubules** (t-tubules) run from the plasma membrane (the sarcolemma) into the centre of the muscle fibre. These are filled with extracellular fluid and therefore are very good for conducting action potentials.
- Within each myofibril, there is a regular parallel arrangement of thick and thin proteins called a **sarcomere** (Figure 4.1).

Sarcomeres

- There can be up to 10,000 sarcomeres per myofibril with each having a resting length of approximately 2μm.
- The length of one sarcomere is the distance between two sets of structural proteins called Z lines.
- A Z line can be found at the midpoint of an I band.
- A set of structural proteins (the M line) can be found at the midpoint of the A band.
- The H zone of a sarcomere is the area in which there are only myosin filaments.
- Actin filaments consist of numerous relatively small molecular weight proteins, whereas myosin filaments consist of numerous relatively large molecular weight proteins.
- The myosin proteins have two distinct sections—the tail and the head. The myosin tails intertwine, whereas the myosin head protrudes. It is the head that will bind to actin during muscular contraction.

Skeletal muscle contraction

- During a typical **isotonic contraction**, the Z lines of the sarcomere move closer together, the lengths of the H zone and the I band are reduced, and the length of the A band remains unchanged.
- These observations suggest that myosin proteins do not move during skeletal muscle contraction whereas actin proteins do move. It also suggests that an interaction between actin and myosin is central to skeletal muscle contraction. This interaction is termed the **sliding-filament mechanism**.
- The sliding-filament mechanism depends on the formation of **cross-bridges** between actin and myosin. Cross-bridges are formed when myosin heads attach to actin.
- Each myosin head contains an actin-binding site and an enzyme which assists in the hydrolysis of adenosine triphosphate (ATP).

- Following innervation, the myosin head attaches to actin and a process called **cross-bridge cycling** begins.
- Cross-bridge cycling is a four-step sequential process which includes attachment of myosin to actin, movement of the cross-bridge, detachment from actin, and hydrolysis of ATP. This process continues until the muscle relaxes.
- A key point, and an important distinction from smooth muscle, is that the myosin heads remain phosphorylated upon relaxation. This indicates that the regulation of skeletal muscle contraction relies on the availability of actin filaments rather than the myosin filaments.

Regulation of skeletal muscle contraction

- The regulation of skeletal muscle contraction relies on the interaction of two additional proteins: **troponin** and **tropomyosin** (Figure 4.2).
- Tropomyosin is a long protein which extends along the length of the actin filament. When the skeletal muscle is relaxed, tropomyosin is located in such a way that it covers the binding sites where the myosin heads would attach.
- Troponin is a small protein which is bound to both actin and tropomyosin. Following stimulation, troponin undergoes a conformational change in shape which results in tropomyosin being moved away from the binding sites on actin so that cross-bridge cycling can occur.

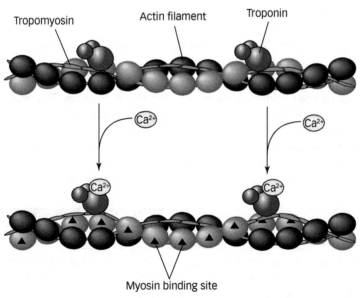

Figure 4.2 Representation of the interaction between actin, myosin, troponin, and tropomyosin

Skeletal muscle

- Calcium plays a key role in the regulatory process. Cytosolic calcium concentration is relatively low; however, following electrical stimulation calcium is moved into the cytosol of the cell. This then binds to troponin, causing the conformational change and the attachment of myosin.
- The process of increasing cytosolic calcium concentration is called **excitation–contraction coupling**.
- In skeletal muscle, the main source of calcium is the sarcoplasmic reticulum. This is another key difference from other types of muscle.

Central activation of skeletal muscle

- A **motor neuron** is a specialized nerve cell with a myelinated sheath that innervates a single skeletal muscle fibre. Because of the myelin sheath, action potentials are propagated along a motor neuron very quickly.
- A **motor unit** consists of a motor neuron and the skeletal muscle fibres it innervates. One motor neuron can innervate numerous skeletal muscle fibres. However, a skeletal muscle fibre is only innervated by one motor neuron.
- The **motor endplate** is an area of the skeletal muscle fibre that is directly under the axon terminal. This area of the muscle fibre has special properties which ensure that when an action potential is generated there it is propagated along the length of the muscle fibre membrane.
- An action potential is generated in the axon of the motor neuron and is propagated quickly to the axon terminal.
- When the action potential arrives at the axon terminal, calcium moves into the axon terminal. Within the axon terminal are vesicles containing acetylcholine. The influx of calcium leads to diffusion of acetylcholine into the neuromuscular junction where it binds to nicotinic receptors on the motor endplate.
- This leads to generation of an action potential in the motor endplate, which is propagated along the length of the sarcolemma. The action potential is propagated into the inside of the muscle fibre via the t-tubules. This then leads to movement of calcium into the cytosol of the cell where it can bind to troponin, which will ultimately lead to skeletal muscle contraction.
- **Acetylcholinesterase** cleaves acetylcholine from the nictonic receptors on the motor endplate, resulting in depolarization of the plasma membrane. This ensures that further innervation and consequent contraction is possible.

Looking for extra marks?

Motor neuron diseases result in the degradation of motor neurons. They are characterized by a progressive loss of ability to contract skeletal muscles. This is due to an inability to increase cytosolic calcium concentration because of the lack of central nervous system activation of the muscle fibre.

Curare is a poison that binds to nicotinic receptors on the motor endplate. Consequently, skeletal muscle contraction is not possible following administration of curare as acetylcholine cannot bind to these receptors and cause depolarization of the motor endplate. This can be lethal because the diaphragm is a skeletal muscle and hence administration of curare has an adverse effect on respiration. Despite this, curare has been used in the past as a muscle relaxant during heart surgery, although it is not used in modern practice.

Acetylcholinesterase inhibitors, such as nerve gases, act by inhibiting acetylcholinesterase. Consequently, the motor endplate remains depolarized and further skeletal muscle contractions are not possible.

Types of muscular contraction

- Skeletal muscle contraction can be described as being isometric or isotonic.
- An **isometric contraction** occurs when force is produced by skeletal muscle but there is no change in muscle length.
- An isometric contraction will occur when the maximum force capacity of a muscle is exceeded. Consequently, force production will be high but there will be no change in the length of the muscle as it cannot move the weight that has been applied to it.
- An isometric contraction does not need to be a maximum voluntary contraction. These types of contractions can also occur at other times. For example, carrying a bag of shopping requires skeletal muscle to produce force but does not necessarily require a change in muscle length.
- An isotonic contraction occurs when force is produced by the muscle fibre and there is a change in muscle length.
- Isotonic contractions (Figure 4.3) are classed as being **concentric contractions** (reduction in length of muscle fibre) or **eccentric contractions** (increase in length of muscle fibre).

Factors affecting skeletal muscle force production

- The amount of force produced by a skeletal muscle fibre depends on the number of cross-bridges that form between actin and myosin (Figure 4.4).
- The **length–tension relationship** shows that there is an optimal overlap, and number of cross-bridges formed, between actin and myosin that results in maximum force production.
- As the length of the sarcomere and the muscle increases, the number of cross-bridges formed is reduced and force production is also reduced.
- Similarly, as the length of the sarcomere and the muscle reduces, the number of cross-bridges formed reduces as a result of excessive overlap between actin and myosin, which again will lead to reduced force production.
- The **force–velocity curve** outlines the relationship between the velocity of muscular contraction and force production. Maximum force production occurs

Skeletal muscle

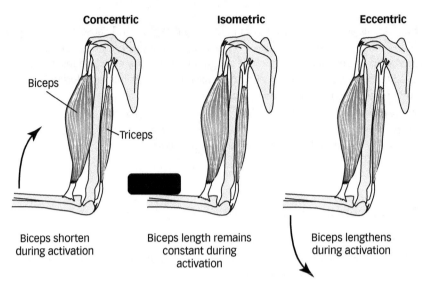

Concentric   **Isometric**   **Eccentric**

Biceps

Triceps

Biceps shorten during activation

Biceps length remains constant during activation

Biceps lengthens during activation

Figure 4.3 Concentric, eccentric, and isometric contractions of the biceps brachii muscle

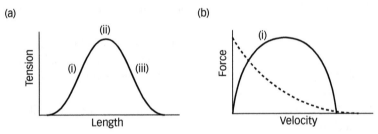

(a)

Tension

(ii)

(i)   (iii)

Length

(b)

Force

(i)

Velocity

Figure 4.4 (a) The length–tension relationship: (i) excessive overlap of actin and myosin; (ii) optimum overlap of actin and myosin filaments; (iii) the effect of increasing muscle length on overlap of actin and myosin filaments. (b) The force–velocity curve and relationship with power: (i) the maximum power production generated at approximately one-third of velocity of shortening

when velocity is zero, i.e. an isometric contraction. As velocity of shortening increases, force is reduced due to an inability to form cross-bridges.

- Power is the product of force and velocity. Maximum power production occurs at approximately one-third of the velocity of shortening.

Types of muscle fibre

- There is some debate about how many different types of muscle fibre exist. However, there are at least three types: type 1, type 2a, and type 2b.
- Type 1 (or slow-twitch) muscle fibres have a relatively small diameter. In addition, they have a high density of mitochondria and a rich blood supply via large numbers of capillaries. These characteristics make this type of muscle fibre relatively slow to fatigue and therefore best suited to aerobic activity.

- Type 2b (or fast-twitch) muscle fibres have a relatively large diameter. They do not have many mitochondria and relatively few capillaries. In addition, they store large amounts of glycogen. Thus these muscle fibres are relatively fast to fatigue and best suited to anaerobic activity.

- Type 2a (or intermediate) muscle fibres combine the properties of type 1 and type 2b fibres. A skeletal muscle will have a relatively small number of this type of fibre, but evidence suggests that they can be trained to become more like type 1 fibres or more like type 2 fibres.

- Every skeletal muscle will have a proportion of these different types of muscle fibre. Some muscle groups, such as postural muscles, will consist predominantly of type 1 muscle fibres as it is beneficial for these muscles to be slow to fatigue. The composition of other muscles varies significantly between individuals. For example, the vastus lateralis muscle may consist of predominantly type 1 fibres or predominantly type 2 fibres. As a result of this, one person may be more suited to participating in exercise that requires endurance (predominantly type 1 fibres) or power (predominantly type 2 fibres). The distribution of muscle fibre type within these muscles appears to be genetically determined.

Looking for extra marks?

Fatigue during high-intensity exercise is a complicated topic. However, the excess hydrogen ions that accumulate following anaerobic breakdown of glucose has historically been thought to play a major role in this process. Recent research has suggested that hydrogen ions only reduce the force production of skeletal muscle fibres when the temperature is very low. This suggests that, under physiological conditions, hydrogen ions play little role in the development of fatigue during high-intensity exercise. The role of inorganic phosphate, one of the products of hydrolysis of phosphocreatine and ATP, in this process has received a lot of attention. This research suggests that inorganic phosphate prevents calcium from leaving the sarcoplasmic reticulum, therefore interfering with excitation–contraction coupling. In addition, inorganic phosphate may prevent the formation of cross-bridges between actin and myosin and therefore force production.

For further reading, see Westerblad H, Allen DG, and Lannergren J (2002) Muscle fatigue: lactic acid or inorganic phosphate the major cause. *News in Physiological Sciences* 17: 17–21.

4.2 SMOOTH MUSCLE

Smooth muscle is found throughout the body and performs numerous important functions. Both the structure and innervation of smooth muscle are different from those of skeletal muscle.

- Unlike skeletal muscle contraction, contraction of smooth muscle is not a voluntary activity. As discussed later, there are three main ways to innervate smooth muscle. However, none of these involve voluntary control.

Smooth muscle

- Smooth muscle plays a vital role in all body systems. Some examples are given here but this should not be considered to be an extensive list.
- Layers of smooth muscle are contained within the tunica media layers of blood vessels. The depth of these smooth muscle layers differs between different vessels (and is described fully in Chapter 5). However, the relaxation or contraction of this smooth muscle leads to an increase or decrease in the diameter of the vessel, which leads to a change in blood flow.
- Smooth muscle surrounds areas of the respiratory system such as the terminal bronchioles. Again, relaxation or contraction of this smooth muscle leads to an increase or decrease in diameter of the bronchiole, which leads to changes in air flow. Asthma attacks are the result of bronchiole smooth muscle contraction and a consequent reduction in air flow.
- Smooth muscle is found in the integumentary system. An example of this is the arrector pili muscle, which is a bundle of smooth muscle fibres attached to a hair follicle. When this contracts in response to a reduction in environmental temperature, hair 'stands on end'.
- Smooth muscle is found extensively in the gastrointestinal system. Layers of smooth muscle surround the stomach, with relaxation causing an increase in stomach volume and contraction causing an increase in intragastric pressure, which leads to movement of foodstuff into the intestine. Layers of smooth muscle are found in all areas of the intestines and contraction of this muscle is central to the process of peristalsis.

Anatomy of smooth muscle

- Unlike skeletal muscle, smooth muscle is not cylindrical. **Smooth muscle fibres** are like sheets that stack on top of each other with no obvious arrangement. They are also relatively short, with a length of up to 200μm and a diameter of up to 10μm.
- Whereas skeletal muscle fibres are multinucleated, smooth muscle fibres are mononucleated. The nucleus of a smooth muscle is found in the middle of the cell.
- Smooth muscle is named as such because no striations are visible when considering the histology of this muscle. As the striations that can be seen in skeletal muscle are due to the arrangement of actin and myosin in sarcomeres, the lack of striations in smooth muscle indicates that no sarcomeres exist in this type of muscle.
- Although there are no sarcomeres in smooth muscle, there are actin and myosin filaments. The proteins have no regular arrangement, but are found at oblique angles. Consequently, when smooth muscle contracts, it not only reduces in length but also twists.
- Smooth muscle fibres do not have Z lines. However, they have **dense bodies**, which perform a similar role to the structural proteins in skeletal muscle.
- Some smooth muscle fibres have **gap junctions** between them. Gap junctions are effectively small pores between the cells which allow movement of ions between

them. Consequently, if an action potential is generated in one smooth muscle fibre, it can be propagated to the next muscle fibre.

- There is sarcoplasmic reticulum within smooth muscle fibres, but there is much less than in skeletal muscle and it is not arranged in any specific manner.

Smooth muscle contraction and regulation

- The presence of actin and myosin within the smooth muscle fibre suggests that smooth muscle contracts in a similar manner to skeletal muscle, i.e. via formation of cross-bridges and a sliding-filament mechanism.
- Smooth muscle contractions can be just as powerful as skeletal muscle contractions. It takes longer to reach maximal force production in smooth muscle because of the slower activity of the enzyme that hydrolyses ATP.
- Regulation of skeletal muscle contraction is the result of interactions between troponin, tropomyosin and actin. However, smooth muscle fibres do not contain troponin. While smooth muscle contraction depends on the interaction between actin and myosin, just as in skeletal muscle, the regulation of this process is different (Figure 4.5).
- The process of excitation–contraction coupling (an increase in cytosolic calcium) in smooth muscle begins as a result of an action potential and ends with the phosphorylation of a myosin head. As stated previously, in skeletal muscle myosin heads are always phosphorylated and excitation–contraction coupling results in the removal of regulatory processes on the actin filament. This is the major difference between the two types of muscle, as the purpose of excitation–contraction coupling in smooth muscle is to phosphorylate the myosin head.
- Following an action potential, calcium is moved into the cytosol of the smooth muscle fibre. Once within the cytosol, calcium binds to **calmodulin** to form a calcium–calmodulin complex. Calmodulin is a protein which is similar to the troponin found in skeletal muscle.

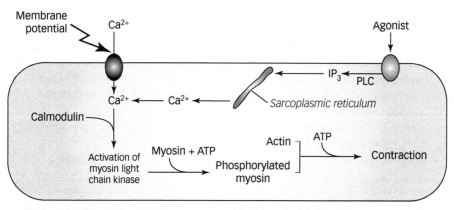

Figure 4.5 Excitation–contraction coupling in smooth muscle fibres

Smooth muscle

- The calcium–calmodulin complex then activates **myosin light chain kinase**, which is an enzyme that assists in the hydrolysis of ATP.
- Following hydrolysis of ATP, the myosin head becomes phosphorylated and cross-bridge cycling can occur.
- Relaxation of smooth muscle occurs when the myosin heads become dephosphorylated.
- In skeletal muscle, the main source of calcium is the sarcoplasmic reticulum. However, as the sarcoplasmic reticulum in smooth muscle fibres is poorly developed and is not arranged in any specific structure, this does not provide an extensive source of calcium in smooth muscle. The extracellular fluid provides a large amount of the calcium required for excitation–contraction coupling in smooth muscle.
- Just like skeletal muscle fibres, smooth muscle fibres exhibit a length–tension relationship, i.e. a reduction in force production away from optimal numbers of cross-bridges. In smooth muscle fibres, the range of lengths is greater than in skeletal muscle fibres.

Innervation of smooth muscle

- Similarly to skeletal muscle, the purpose of smooth muscle innervation is to increase cytosolic calcium concentration.
- The three main mechanisms of smooth muscle innervation are autonomic nervous system activity or hormones, **pacemaker potentials**, and local factors.
- The sarcolemma of smooth muscle fibres contains receptors for a variety of neurotransmitters secreted from **varicocities** of neurons of the autonomic nervous system. Activation of these receptors could lead to an increase or decrease in cytosolic calcium. Consequently, central nervous system activity can cause contraction or relaxation of smooth muscle. This is unlike skeletal muscle, which will only contract in response to central nervous system activity.
- The neurotransmitters most commonly associated with smooth muscle activity are acetylcholine (sympathetic) and noradrenaline (parasympathetic). In practice, it is the receptor and not the neurotransmitter which determines whether there will be an increase or decrease in cytosolic calcium. For example, noadrenaline may bind to alpha-adrenergic receptors and cause an increase in cytosolic calcium concentration (causing contraction of smooth muscle), or it may bind to beta-adrenergic receptors and cause a reduction in cytosolic calcium concentration (causing relaxation of smooth muscle). Consequently, administration of noradrenaline may have different effects on different body systems depending on the predominant receptor that is found.
- Some smooth muscle fibres exhibit pacemaker potentials. These fibres have unstable resting membrane potentials due to the movement of ions across the membrane, which favours a gradual increase in resting membrane potential. When the membrane potential reaches its threshold value, an action potential is produced. Owing to the existence of gap junctions between some smooth muscle

fibres, it is possible for sections of smooth muscle to contract without any central nervous system activity as a result of these pacemaker potentials.

- A variety of local factors can have a direct effect on smooth muscle contraction or relaxation. Examples of these local factors include partial pressures of oxygen and carbon dioxide, hydrogen ions, ATP, ADP, nitric oxide, and bradykinin. These local factors have a direct effect on the movement of calcium into the smooth muscle fibre and therefore whether it will contract or relax. For example, exercise results in a reduction in the partial pressure of oxygen and ATP as well as an increase in the partial pressure of carbon dioxide, hydrogen ions, ADP, nitric oxide, and bradykinin. These local factors lead to a reduction in calcium entering vascular smooth muscle fibres, which leads to relaxation, a reduction in vessel diameter, and an increase in blood flow to exercising skeletal muscle.

Types of smooth muscle

- It is very difficult to classify smooth muscle into different types. However, it can be classed broadly as being multi-unit or single-unit.
- As the name implies, **multi-unit smooth muscle** does not contract as a whole muscle unit. Instead, the muscle contracts in individual units. Therefore multi-unit smooth muscle tends to have relatively few gap junctions and is innervated almost exclusively by autonomic nervous system activity or circulating hormones.
- Again as the name implies, **single-unit smooth muscle** contracts as a single unit. This is due to the existence of a large number of gap junctions. Single-unit smooth muscle can be innervated by the autonomic nervous system, circulating hormones, local factors, and pacemaker potentials.

4.3 CARDIAC MUSCLE

Cardiac muscle is found exclusively in the myocardial layer of the heart. Its contraction leads to an increase in pressure within the heart which ultimately leads to movement of blood through the chambers of the heart or into the pulmonary or systemic circulations.

- **Cardiac muscle fibres** are relatively short, with a length up to 100µm and diameter up to 20µm.
- Like smooth muscle fibres, cardiac muscle fibres are mononucleated with the nucleus in the middle of the cell.
- Cardiac muscle fibres branch and contain gap junctions for transfer of action potentials.
- Like skeletal muscle, cardiac muscle is classed as being striated. Again, this is due to the existence of myofibrils and sarcomeres. Cardiac muscle fibres have t-tubules for propagation of action potentials just like skeletal muscle.
- Cardiac muscle has reasonably well-developed sarcoplasmic reticulum. There is not as much as in skeletal muscle but more than is found in smooth muscle.

Cardiac muscle contraction, regulation, and innervation

- The existence of actin and myosin in sarcomeres indicates that the contraction of cardiac muscle occurs as a result of a sliding-filament mechanism, just as in skeletal and smooth muscle.
- The regulation of cardiac muscle contraction is the same as in skeletal muscle, i.e. the interaction between troponin, tropomyosin, and actin.
- The purpose of excitation–contraction coupling is the same as in the other muscle types, i.e. to cause an increase in cytosolic calcium concentration.
- As the sarcoplasmic reticulum in cardiac muscle fibres is not as well developed as in skeletal muscle, calcium is also derived from extracellular fluid.
- Following the arrival of an action potential, some calcium is moved from the extracellular fluid into the cardiac muscle fibre and binds to troponin, causing cross-bridge cycling to occur. Some of the calcium from the extracellular fluid also enters the sarcoplasmic reticulum. This causes more calcium to move from the sarcoplasmic reticulum into the cytosol of the cardiac muscle fibre. This results in an augmentation of excitation–contraction coupling.
- Innervation of cardiac muscle fibres comes from the **conduction system** of the heart. This is described in detail in Chapter 5. Pacemaker potentials are generated within the **sinoatrial node** and propagated through the conduction system to cardiac muscle fibres, causing increases in cytosolic calcium concentration and muscle contraction. The existence of gap junctions means that it is not necessary for each cardiac muscle fibre to be innervated directly.

Check your understanding

List the major differences in structure and innervation between the three types of muscle. (*Hint: consider the anatomy of the types of muscle and whether contraction is voluntary or involuntary*)

What are the major differences in regulation of contraction between skeletal and smooth muscle? (*Hint: consider the site of regulation in each type of muscle*)

Outline the main relationships that affect force production in skeletal muscle. (*Hint: see Figure 4.4*)

5 Cardiovascular physiology

The human body contains approximately 75 trillion cells. These cells require oxygen, nutrients, and an environment conducive to normal cellular function in order to work properly. The cardiovascular system provides cells with oxygen and nutrients, as well as transporting metabolic waste products away from cells. It consists of the heart and an extensive system of vessels known as the circulatory system. The heart acts as a pump while the circulatory system acts as a transport system for blood.

Key concepts

- Normal functioning of the cardiovascular system is essential in order to maintain an appropriate cellular environment for all cells in the human body.
- The heart can be thought of as a pump that ejects blood into two different circulations: the pulmonary and systemic circulations.
- Changes in pressure throughout the cardiovascular system are essential to ensure an appropriate flow of blood.
- The cardiovascular system is capable of adapting quickly to various stimuli in order to ensure an adequate blood flow to cells.
- Arterioles are the main site of regulation of mean arterial pressure and effectively control blood flow to tissues.

continued

- Cardiac output is the product of stroke volume and heart rate. The main determinant of stroke volume is end-diastolic volume, which is influenced primarily by venous return.
- The autorhythmic nature of cardiac muscle contraction is the result of the action of the conduction system of the heart.
- An electrocardiogram can be used to examine the electrical activity of the heart.

5.1 HEART STRUCTURE AND FUNCTION

The heart is roughly the size of a clenched fist and is located to the **anterior** (front) of the chest wall in the upper left quadrant of the chest. An understanding of the structure of the heart aids in understanding its function.

- A fluid-filled sac, called the **pericardium**, surrounds the heart.
- The fluid within the pericardium serves as a shock absorber for the heart.
- The pericardium consists of two layers of connective tissue: a serous (inner) layer and fibrous (outer) layer.
- The fibrous layer of the pericardium is attached to the diaphragm, which accounts for the movement of the heart during breathing, i.e. contraction of the diaphragm.
- **Cardiac tamponade** occurs when the fibrous layer is ruptured, causing an increase in the amount of fluid within the pericardium, which results in an increase in pressure acting on the heart muscle.

Structure

- The heart wall consists of three layers: the **epicardium**, the **myocardium**, and the **endocardium**.
- The epicardium is a layer of connective tissue on the outside of the heart which attaches to the serous layer of the pericardium.
- The myocardium is the thickest layer of the heart and consists of cardiac muscle, blood vessels, and nerve fibres.
- The endocardium is a layer of connective tissue and endothelial cells on the inside of the heart.

Blood supply

- As with all cells, cardiac tissue requires a blood supply.
- Blood is supplied by the left and right **coronary arteries**, which supply the left and right sides of the heart, respectively.
- Occlusion of the coronary arteries results in reduced blood flow to the heart, which can result in conditions such as **angina pectoris** and/or **myocardial infarction**.
- The extent and duration of the occlusion is important when determining the extent of damage to cardiac tissue that may have occurred.

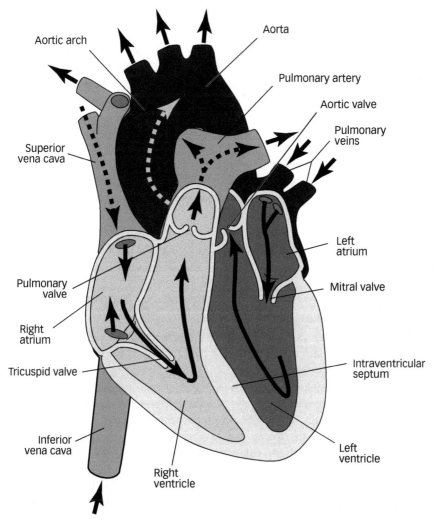

Figure 5.1 Schematic representation of the heart including the mitral and tricuspid valves

Anatomy

- The heart consists of four main chambers (Figure 5.1).
- The left and right **atria** are situated in the upper left and right portions of the heart while the left and right **ventricles** are situated in the lower left and right portions of the heart.
- The left and right atria are separated by a thin muscular wall called the **interatrial septum** while the left and right ventricles are separated by the **interventricular septum**. The purpose of these walls is to ensure that blood cannot move between the two atria or the two ventricles.

- The left atrium and left ventricle serve as a single pump which drains into the aorta and thus provides blood to the **systemic circulation**.
- The right atrium and right ventricle serve as a single pump which drains into the pulmonary artery and thus provides blood to the **pulmonary circulation**.
- Blood cannot move freely between the left atrium and left ventricle, or the right atrium and right ventricle, because of the existence of valves.

Heart valves

- Heart valves ensure that blood can only move in one direction.
- The valve between the right atrium and right ventricle is called the **tricuspid valve**.
- The valve between the left atrium and left ventricle is called the **bicuspid valve** or the mitral valve.
- Mitral regurgitation is a condition where the mitral valve does not close properly, resulting in blood returning from the left ventricle to the left atrium.
- **Papillary muscles** are attached to the valves and to the ventricle walls via **chordae tendinae**.
- Valves open in response to changes in pressure and, in particular, differences in pressure between the atria and the ventricles.
- When pressure within the atria exceeds the pressure within the ventricles, which occurs during contraction of the atria, the valves will open and thus blood will empty into the ventricles.
- The **pulmonary valve** is the valve between the right ventricle and the pulmonary artery, and the **aortic valve** is the valve between the left ventricle and the aorta. These valves control the flow of blood from the ventricles into the pulmonary and systemic circulation.
- The pulmonary and aortic valves also open in response to pressure differences between the ventricle and the vessel, i.e when pressure within the ventricle exceeds the pressure within the vessel, the valve will open and allow blood to move into the circulatory system. This increase in ventricular pressure is due to contraction of the ventricle.

5.2 THE CIRCULATORY SYSTEM

Following ejection from the ventricles, blood enters the circulatory system. The circulatory system is a series of vessels, which vary in elements of their structure, that carry blood either to the lungs (in the case of the pulmonary circulation) or to the rest of the body (in the case of the systemic circulation).

- The circulatory system consists of **arteries, arterioles, capillaries, venules,** and **veins.**
- Arteries and arterioles carry blood away from the heart and thus can be likened to efferent neurons of the nervous system (see Chapter 2).
- Venules and veins carry blood back to the heart and thus can be likened to afferent neurons of the nervous system (see Chapter 2).

- Capillaries perform the ultimate role of the cardiovascular system as they are the only site where gas exchange can occur.

Structure

- All elements of the circulatory system have a similar structure (Figure 5.2).
- All elements contain a **tunica intima**, a **tunica media**, and a **tunica externa**. The thickness of each layer differs amongst the various vessels of the circulatory system.
- The tunica intima consists of a thin layer of endothelium and some connective tissue. The amount of connective tissue depends on the vessel. In arteries, the tunica intima also contains a thin layer of elastic fibres called the **internal elastic membrane**.
- The tunica media consists almost entirely of smooth muscle. Again, the amount of smooth muscle varies between different vessels. The contraction or relaxation of the smooth muscle in this layer ultimately determines the diameter of the vessel and, consequently, the rate at which blood flows through it. In arteries, there is a further thin layer of elastic fibres called the **external elastic membrane.**
- Together, the internal and external elastic membranes in arteries allow these vessels to withstand the large amount of pressure that is exerted on their walls when blood is ejected from the heart. This also assists with movement of blood from arteries into arterioles owing to a reduction in vessel diameter when the elastic layers begin to return to their resting state.
- The tunica externa largely consists of connective tissue with the amount depending on the vessel in question. This connective tissue acts as an anchor and stabilizes the vessel.
- The thickness of the vessel wall is greater in arteries than in veins. This is largely due to the greater thickness of the tunica media in arteries, even though the tunica externa is larger in veins.

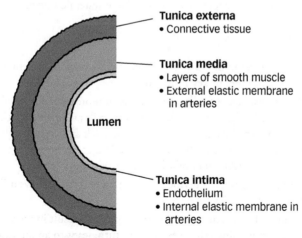

Figure 5.2 Layers of the circulatory system

- The diameter of the lumen is smaller in arteries than in veins when no blood flow is present. This is due to elastic recoil of arteries via the internal and external elastic membranes, which are only present in these vessels.
- No gas exchange occurs in arteries, arterioles, venules, or veins because of the thickness of the vessel wall. Capillaries have little or no tunica media or externa; thus the vessel wall is very thin, which allows gases to diffuse easily.

The path of blood

- Deoxygenated blood is returned to the right atrium. As a result of an increase in pressure within the right atrium, the tricuspid valve opens and blood is emptied into the right ventricle.
- Pressure inside the right ventricle increases due to increases in the volume of blood within the ventricle and ventricular contraction, the pulmonary valve opens, and blood is ejected into the pulmonary artery.
- From the pulmonary artery, blood moves into the pulmonary arterioles and into the pulmonary capillaries surrounding the lungs.
- Blood becomes oxygenated by diffusion of oxygen from alveoli across capillary walls.
- Oxygenated blood then enters the pulmonary venules before moving to the pulmonary veins, which drain into the left atrium.
- Following an increase in pressure, the bicuspid valve opens and allows blood to drain into the left ventricle.
- Pressure within the left ventricle increases and the aortic valve opens allowing blood to be ejected into the aorta (a large systemic artery).
- Blood is then moved into systemic arterioles and capillaries where oxygen diffuses across capillary walls and becomes deoxygenated at the site of active tissue.
- Blood is then moved into systemic venules and veins. It is returned to the right atrium via the inferior or superior vena cava (large systemic veins).

The endothelium

- The tunica intima layer in blood vessels contains a layer of endothelium. This layer of cells has a number of key functions.
- As it is the innermost layer of cells, the endothelium provides a level of protection to the blood vessel. Under normal circumstances, cells are unable to stick to the endothelium; however, if damage occurs cells can stick and cause plaques. This is consistent with the 'response to injury' hypothesis that has been suggested as the main reason for the development of **atherosclerosis.**
- The layer of endothelium has a permeable membrane, which allows diffusion of gases and fluids in capillaries.
- The endothelium releases a number of compounds that result in relaxation of smooth muscle surrounding arterioles. This, in turn, leads to an increase in blood flow through the arteriole.

- The process of **angiogenesis**, or new capillary growth, is mediated via the endothelium.
- The endothelium is responsible for secretion of a number of growth factors, which are involved in immune response and numerous substances that are central to the process of blood clotting.

Arteries

- Arteries have relatively thick muscular walls due to the presence of large amounts of smooth muscle in the tunica media.
- The smooth muscle in the tunica media contracts or relaxes largely in response to input from the sympathetic nervous system.
- Contraction of arterial smooth muscle leads to a reduction in vessel diameter (**vasoconstriction**) and relaxation leads to an increase in vessel diameter (**vasodilation**).
- Arteries can be classified as muscular or elastic arteries.
- Elastic arteries are usually relatively large vessels such as the aorta. The internal and external elastic membranes are well developed and allow a large amount of pressure to be absorbed.
- Muscular arteries have relatively large amounts of smooth muscle in the tunica media. They tend to be responsible for the distribution of blood flow to various organs. An example of a muscular artery is the carotid artery, which is often used when taking a person's pulse.

Arterioles

- Arterioles have a very small diameter and may have very small amounts of smooth muscle or several layers of smooth muscle in the tunica media.
- Arterioles are also known as resistance vessels as they are very important in determining blood flow to tissues.
- Like arteries, arterioles can vasoconstrict or vasodilate as a result of smooth muscle contraction or relaxation (Figure 5.3).
- Arteriolar smooth muscle contracts in response to central nervous system activation, as well as a number of local factors.
- As arterioles are very important in determining blood flow, they have a major impact on **blood pressure**.

Capillaries

- Capillaries are the only site of gas exchange. This is possible because the vessel wall is very thin and often only consists of a thin layer of tunica intima.
- The thin walls of capillaries coupled with the short distance between the capillary and the organ means that gas exchange can occur very rapidly.

The circulatory system

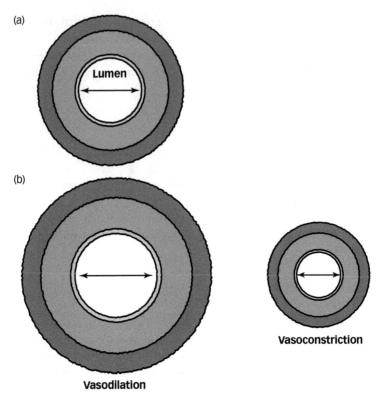

Figure 5.3 (a) Arteriolar diameter under resting conditions and (b) the effect of smooth muscle contraction and relaxation on vessel diameter

- One arteriole can give rise to numerous capillaries, which form a bed around an organ. Blood flow rate decreases as the cross-sectional area of the vessel increases. As there are a large number of capillaries, the cross-sectional area is high and flow rate is relatively slow. This allows the maximum amount of time possible for gas exchange.
- Capillaries do not act as a single unit. At the entrance to each capillary, there is a pre-capillary sphincter which can contract or relax in response to various stimuli and lead to reduced or increased blood flow through that capillary.
- Capillaries are classed as being either continuous or fenestrated.
- Continuous capillaries have a single continuous layer of endothelium. As a result, it is not possible for proteins to cross these capillaries. However gas, water, and some other solutes can.
- Fenestrated capillaries have small pores in the layer of endothelium which allow rapid exchange of proteins. The liver has a large amount of fenestrated capillaries.

Venules and veins

- Each capillary drains into numerous venules. Venules can sometimes be difficult to distinguish from capillaries, as they have a very small diameter and very poorly developed tunica media and externa.

- Veins can be classed as medium-sized or large-sized.
- Medium-sized veins have a very small tunica media layer with little smooth muscle.
- Large-sized veins have a more developed tunic media layer with large amounts of smooth muscle, which can contract or relax in response to central nervous system activation.
- The flow of blood back to the heart is assisted by the presence of valves. Blood pressure in veins is relatively low compared with that in arteries. In addition to a relatively small pressure difference assisting blood flow, the blood has to overcome the effect of gravity to return to the heart. Valves ensure unidirectional movement of blood, which prevents pooling of blood in the lower extremities. Problems with these valves can lead to conditions such as varicose veins.
- At any one time, approximately 60% of all blood is in the venous circulation.

5.3 BLOOD

As previously described, the main function of the cardiovascular system is to provide oxygen, nutrients, and an appropriate environment for cells. Therefore blood is an essential component of the cardiovascular system.

- Blood can be thought of as a collection of cells or cell fragments within a fluid compartment.
- One of its main functions is the transport of gases, nutrients, and hormones. In the case of gases, oxygen and carbon dioxide are transported around the body. Nutrients are absorbed from the gastrointestinal tract and transported to storage sites freely or bound to proteins. Hormones are produced in a variety of organs and transported in the blood to elicit an action elsewhere.
- Blood is strongly involved in regulation of acid–base balance, as well as immune function and regulation of body temperature.
- The cells present in blood are **erythrocytes** (red blood cells) and **leukocytes** (white blood cells).
- The cell fragments consist largely of **platelets**.
- The cells and cell fragments are suspended in plasma. In addition to the cells and cell fragments, plasma contains numerous proteins and electrolytes. The main protein present in plasma is **albumin**. Globulins and clotting factors are also present.
- Once collected, blood samples are often centrifuged to separate blood into its constituent parts. When this is performed, the percentage of red blood cells (haematocrit) can be calculated and plasma can be obtained for measurement of a variety of factors. A very thin layer of white blood cells can also be seen between the red blood cells and the plasma. Some researchers may analyse serum instead of plasma.

Erythrocytes

- Erythrocytes, or red blood cells, make up approximately 99.9% of all cells in the blood. They contain **haemoglobin**, which accounts for the red colour of blood.

Blood

- Erythrocytes are formed via erythropoiesis in the bone marrow. A hormone called **erythropoietin** stimulates the bone marrow to produce erythrocytes. This hormone has often been abused by athletes as it stimulates red blood cell production and increases oxygen-carrying capacity.
- Erythrocytes have a biconcave disc structure, which enables the cell to be flexible. This structure also ensures that erythrocytes have a relatively high surface area, which is important for gas diffusion.
- Erythrocytes do not have a nucleus, mitochondria, or organelles. As a result of this, each erythrocyte has a lifespan of approximately 120 days.
- The lack of mitochondria means that energy has to be obtained from the anaerobic breakdown of glucose from the plasma. This means that the oxygen that the erythrocyte is carrying is not used by the erythrocyte itself.

Haemoglobin

- Haemoglobin is a complex protein consisting of four chains of amino acids (Figure 5.4).
- Each amino acid chain contains a heme molecule, which includes iron. Oxygen binds to the iron that is present in this heme molecule.
- There is a very weak association between oxygen and haemoglobin. Consequently, there can be very fast release of oxygen at active tissues.
- Each haemoglobin molecule can transport four oxygen molecules. When oxygen is bound to haemoglobin it is called **oxyhaemoglobin** whereas when carbon dioxide is bound it is called **carboxyhaemoglobin**.

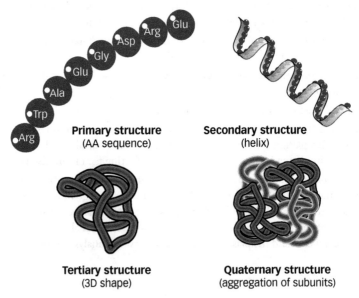

Primary structure
(AA sequence)

Secondary structure
(helix)

Tertiary structure
(3D shape)

Quaternary structure
(aggregation of subunits)

Figure 5.4 Haemoglobin structure

- The exchange of oxygen and carbon dioxide depends primarily on the plasma levels of the appropriate gas.
- When plasma oxygen levels are low and carbon dioxide levels are high (as occurs around active tissue), oxygen is released from haemoglobin and carbon dioxide is bound.
- When plasma oxygen levels are high and carbon dioxide levels are low (as is the case around the lungs), carbon dioxide is released from haemoglobin and oxygen is bound.

5.4 BLOOD PRESSURE

Blood pressure is often thought of in terms of clinical conditions such as hypertension. However, it is important to remember that blood flow relies on pressure gradients, so an understanding of how these pressure gradients are created is very important.

- The pressure exerted on the walls of a vessel depends on the volume of blood in that vessel and the **compliance** of the vessel's walls.
- In an artery, pressure is determined by the amount of blood exiting the ventricle (stroke volume) and the compliance of the vessel.
- While blood is exiting the heart, pressure in the artery increases. This peaks at the end of **systole** (the ejection phase of the **cardiac cycle**) and is called **systolic pressure**.
- As the pressure inside an arteriole is less than in an artery, blood moves from the artery into the arteriole. This process is assisted by elastic recoil of the artery.
- Only about a third of the volume of blood ejected from the heart moves into the arterioles. The rest stays in the arteries. As a result, there is always pressure in the artery. The lowest pressure occurs at the end of **diastole** (the filling phase of the cardiac cycle) and is called **diastolic pressure**.
- Blood pressure is normally presented as systolic pressure/diastolic pressure and is recorded in millimetres of mercury (mmHg).
- A typical healthy blood pressure is 120/80mmHg.
- It is important to remember that this is the pressure in the systemic circulation. Pressures in the pulmonary circulation are much lower.
- Pulse pressure is calculated by subtracting the diastolic pressure from the systolic pressure, i.e. 120 − 80 = 40mmHg in a typical healthy person.
- **Mean arterial pressure** is often used as an indicator of the average pressure across the entire system. As the filling phase of the cardiac cycle (diastole) is longer than the ejection phase of the cardiac cycle (systole), mean arterial pressure is not simply the average of systolic and diastolic pressures. Instead, it is calculated using the following formula:

mean arterial pressure = diastolic pressure + (pulse pressure)/3

Blood pressure

- Thus a person with a blood pressure of 120/80mmHg, has a mean arterial pressure of 93mmHg.
- Pressure inside arteries exceeds that in arterioles which exceeds that in capillaries which exceeds that in venules which exceeds that in veins. Pressure in veins can be as low as 10–15mmHg. However, it is the pressure gradient between the various vessels that ensures flow of blood through the circulatory system.

Measurement of blood pressure

- Blood pressure of the systemic circulation is usually measured with a **sphygmomanometer**.
- The cuff is placed around the upper arm and inflated to a pressure that exceeds systolic pressure.
- As a result of this, there is no blood flow through the brachial artery. If a stethoscope is placed over the site, no sounds will be heard.
- The pressure in the cuff is reduced slowly. At the point when systolic pressure is achieved, blood flow through the brachial artery will return and generate distinctive sounds. As the pressure in the cuff continues to be reduced, more sounds can be heard until the point of diastolic pressure.
- The sounds that can be heard throughout this process are called **Korotkoff sounds**.

Looking for extra marks?

Essential hypertension is classed as a cardiovascular disease and is a chronic elevation in blood pressure. Because of the chronic increase in pressure, the cardiac muscle in the myocardium of the left ventricle hypertrophies. This causes an increase in left ventricular mass. This is very different from the increase in left ventricular mass that is seen as a result of aerobic exercise training programmes (see Chapter 10). Blood supply to the myocardium has to be increased because of the increase in left ventricular mass and, over time, this can lead to a variety of problems including myocardial infarction and stroke. A variety of organs can be affected by hypertension, including the kidneys. Untreated hypertension can lead to serious problems in kidney function which will eventually lead to kidney damage and/or end stage renal failure. A systemic blood pressure of up to 140/90mmHg is considered as 'pre-hypertension'; further increases are denoted stage 1 (140–160/90–100mmHg), stage 2 (160–180/100–110mmHg), stage 3 (180–210/110–120mmHg), and stage 4 (>210/>120mmHg) hypertension. The cause of hypertension is a matter of some debate. However, lifestyle factors such as diet and/or physical activity seem to be important. There appears to be a significant relationship between obesity and hypertension, and this is covered further in Chapter 10. A lot of research has focused on the role of salt, more specifically sodium, in the development of hypertension. Excessive salt intake is likely to lead to increases in circulating blood volume and water retention. In turn, this is likely to increase blood pressure. Sedentary individuals are more likely to become hypertensive than physically active individuals. Changes in central nervous system activation and peripheral resistance following a period of aerobic exercise training result in reductions in blood pressure. Initially, weight loss or changes in diet (including reduced saturated fat and salt intake) can be used as a treatment of hypertension. A programme of physical activity may also provide beneficial effects; however, this should be of low to medium intensity rather than high intensity. If blood pressure remains elevated, a variety of pharmacological aids can be used. Amongst the most commonly used are angiotensin-converting enzyme (ACE) inhibitors. ACE is an important component of the renin–angiotensin system, which has a major role in regulating blood pressure. By inhibiting ACE, water retention and vasoconstriction is minimized, resulting in a reduction in blood pressure. ACE inhibitors tend to be administered over the long term and are not suitable for some populations.

Pressure and flow through arterioles

- Poiseuille's law states that flow is proportional to the pressure gradient divided by the resistance to flow:

$$F = \Delta P / R$$

- There is a pressure gradient between arteries and arterioles, which ensures that blood flows down the vascular tree. However, the resistance of arterioles determines the rate at which that flow occurs. For this reason, arterioles are sometimes known as 'resistance vessels'.

- If resistance in the arteriole increases, blood flow through the arteriole reduces, whereas if resistance in the arteriole decreases, blood flow through the arteriole increases.

- The resistance of the arteriole is altered by changes in its diameter, i.e. vasoconstriction or vasodilation.

- The change in diameter of arterioles is the result of contraction or relaxation of the smooth muscle surrounding that arteriole, i.e. if the smooth muscle contracts, the diameter of the arteriole reduces, resistance is increased, and blood flow is reduced.

- Given that arterioles are transporting oxygenated blood to capillaries, which will supply organs, they effectively control blood flow and oxygen delivery to organs. In addition, they have a major impact on mean arterial pressure.

- Arteriolar smooth muscle relaxes or contracts in response to local or extrinsic factors.

- Local factors that affect blood flow include **active hyperaemia** and **autoregulation**.

- The process of active hyperaemia occurs as a result of an increase in a number of metabolites due to activity within a certain tissue. Increases in partial pressure of carbon dioxide, hydrogen ions, bradykinin, and nitric oxide, as well as reductions in the partial pressure of oxygen, are some of the variables that are affected by an increase in metabolism. These metabolic by-products act directly on arteriolar smooth muscle and cause relaxation, which, in turn, leads to an increase in blood flow. Thus, active hyperaemia is an important mechanism for ensuring adequate blood flow to tissues where metabolic demand is high.

- Autoregulation occurs as a result of changes in pressure within the arteriole. When pressure inside the arteriole is reduced (perhaps as a result of a small blockage) but metabolic activity remains constant, an increase in metabolic by-products is observed. This results in vasodilation via the same mechanism as for active hyperaemia. Consequently, it is the event that distinguishes autoregulation from active hyperaemia, although the mechanism for increasing blood flow is the same.
- Local controls of blood flow exist to ensure adequate blood supply to organs; extrinsic control of blood flow ensures regulation of mean arterial pressure.
- Extrinsic control of blood flow includes activation from the central nervous system and circulating hormones.
- Most tissues receive parasympathetic and sympathetic innervations. However, arterioles do not. They receive very little parasympathetic innervation but extensive sympathetic innervation.
- Increased sympathetic tone leads to contraction of arteriolar smooth muscle, vasoconstriction, an increase in pressure, and a reduction in blood flow. Noradrenaline is used as a neurotransmitter.
- Arteriolar smooth muscle relaxation and vasodilation occur as a result of a reduction in sympathetic tone rather than an increase in parasympathetic tone.
- A number of the metabolites mentioned so far, which are produced as a result of performing physical activity, lead to an increase in sympathetic nervous system activation and a consequent increase in blood pressure. This is termed the exercise pressor response.
- Increases in circulating concentrations of adrenaline, angiotensin II, and vasopressin tend to lead to arteriolar smooth muscle contraction and vasoconstriction. However, adrenaline can cause arteriolar smooth muscle relaxation if it binds to the appropriate receptor. Increases in circulating concentrations of atrial natriuretic factor (sometimes referred to as atrial natriuretic peptide) lead to arteriolar smooth muscle relaxation, vasodilation, and a reduction in blood pressure.

Pressure and flow through veins

- At any one time, approximately 60% of all blood is in systemic veins but blood pressure is very low. This is due to the large diameter of the vessels and the compliant nature of the vessel walls.
- Gravity has to be overcome in addition to the small pressure gradient that exists to enable blood flow from the veins to the right atrium.
- An increase in venous pressure leads to an increase in blood flow through the vein. This increase in venous pressure is due to venous smooth muscle contraction.

- Venous smooth muscle contraction is enabled by an increase in central nervous system tone.
- Importantly, an increase in sympathetic nervous system activity leads to an increased arteriolar and venous smooth muscle contraction. This results in a reduction in blood flow through arterioles but an increase in blood flow through veins.
- In addition to sympathetic nervous system activation, blood flow through veins is increased as a result of **skeletal muscle pumps**. The contraction of skeletal muscle results in a reduction in venous diameter and an increase in pressure. Similarly, the **respiratory muscle pump** also leads to an increase in venous blood flow.

Regulation of mean arterial pressure

- **Cardiac output** is defined as the volume of blood ejected from the ventricle in 1 minute and is the product of **heart rate** (the number of myocardial contractions per minute) and **stroke volume** (the volume of blood emptied from the ventricle per myocardial contraction).
- **Total peripheral resistance** is the resistance of systemic vessels to blood flow.
- Mean arterial pressure can be calculated by multiplying cardiac output by total peripheral resistance.
- Consequently, any factor that increases heart rate, stroke volume, or total peripheral resistance will increase mean arterial pressure.
- The main factor that affects stroke volume is **venous return** (as this affects **end-diastolic volume** as described later). Venous return is determined by venous pressure (which is determined by sympathetic nervous system activation) as well as by skeletal and respiratory muscle pumps.
- The main factor that affects heart rate is central nervous system activation (as described later).
- Arteriolar smooth muscle contraction provides the main drive for total peripheral resistance and this is affected by sympathetic nervous system activation, circulating hormones, and some local factors (as described earlier).

5.5 THE CONDUCTION SYSTEM

The properties of cardiac muscle are covered in Chapter 4. One of the key properties that distinguishes cardiac muscle from skeletal and smooth muscle is its ability to contract without any need for central nervous system input. This is achieved through the properties of the conduction system (Figure 5.5).

Sinoatrial node

- The **sinoatrial node** is a small group of specialized cells located in the upper right-hand corner of the right atrium.
- Cells in the sinoatrial node are amongst the few cells in the human body that are capable of producing pacemaker potentials.

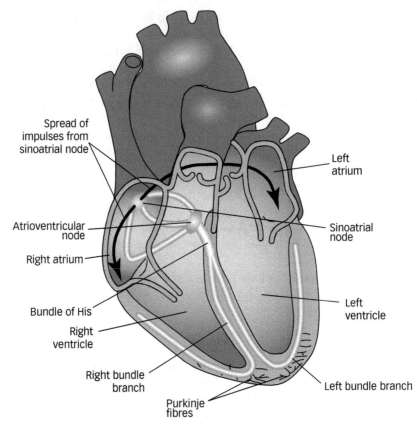

Figure 5.5 The conduction system of the heart including the sinoatrial node, atrioventricular node, bundles of His, and Purkinje fibres

- Unlike most cells in the body, the cells within the sinoatrial node do not have a stable resting membrane potential.
- As the membrane of these cells is porous, ions can easily be exchanged, which leads to a slow increase in resting membrane potential until this reaches the threshold value and an action potential is produced.
- Gap junctions are specialized connections between cells that allow the movement of ions, and therefore action potentials, between cells.
- The presence of gap junctions between cardiac muscle cells means that once a pacemaker potential is generated in the sinoatrial node, an action potential can be propagated to all cardiac muscle resulting in a coordinated contraction of the atria.
- Without any central nervous system input, the sinoatrial node would produce about 100 action potentials per minute, leading to a heart rate of 100 beats per minute.
- At rest, the heart is constantly under parasympathetic influence from the vagus nerve. This reduces the number of action potentials produced by the sinoatrial node to 60–70 per minute. This is termed vagal restraint.

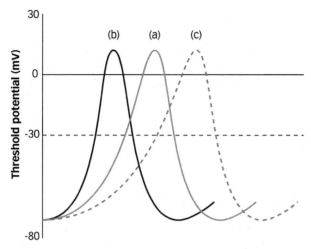

Figure 5.6 A pacemaker potential (a) without central nervous system innervation,
(b) with sympathetic innervation, and (c) with parasympathetic innervation

- When the heart rate is required to increase (e.g. during exercise) the parasympathetic influence on the sinoatrial node is gradually removed. Following this, sympathetic activation increases the number of action potentials produced per minute by increasing the rate at which pacemaker potentials are produced in the sinoatrial node (Figure 5.6).
- A period of aerobic exercise training results in numerous physiological adaptations, including increased parasympathetic influence on the sinoatrial node at rest, and results in a reduced resting heart rate. This reduction in resting heart rate is achieved as a result of the increased resting stroke volume that occurs following exercise training because of left ventricular cardiac muscle hypertrophy.

Atrioventricular node

- The **atrioventricular node** is a small group of cells in the bottom right-hand corner of the right atrium.
- The atrioventricular node acts as a link between the atria and the ventricles.
- Following depolarization of the atria, the atrioventricular node briefly delays the action potential before it is propagated to the ventricles.
- This ensures that contraction of the atria is completed before contraction of the ventricles begins, thereby maximizing the efficiency of heart contraction.

Bundles of His and Purkinje fibres

- The **bundles of His** are specialized cardiac cells that allow propagation of an action potential from the atrioventricular node to the left and right ventricles.
- The left and right bundles of His extend from the atrioventricular node down the left and right sides of the interventricular septum to the bottom of each ventricle

before branching into **Purkinje fibres**, which innervate ventricular cardiac muscle cells within the myocardium and ultimately result in ventricular contraction.

- Ventricular contraction is made more efficient by the extension of the bundles of His towards the base of the ventricles before branching into Purkinje fibres. This results in ventricular contraction beginning at the base of the ventricle and extending upwards. This maximizes movement of blood into the pulmonary and systemic circulations.
- Propagation of action potentials through the ventricles is very fast, taking less than a tenth of a second.

5.6 THE ELECTROCARDIOGRAM

The **electrocardiogram** (ECG) is a depiction of the electrical events of the heart and is commonly used in clinical settings to determine potential problems with the conduction system that may result in arrhythmias.

- Action potentials produced in the myocardium are recorded via electrodes placed on the skin.
- A simple ECG trace can be produced using three electrodes that form a triangle across the heart. However, in clinical settings, nine or twelve electrodes may be used to increase the accuracy of readings.
- An ECG does not depict changes in cardiac muscle membrane potential or the contraction of the myocardium itself. It does depict the electrical events that lead to the contraction of cardiac muscle.

Events of the ECG

- The standard ECG trace shows several clear events. These are the P-wave, the QRS complex, and the T-wave (Figure 5.7).
- The P-wave represents depolarization of the atria, which will ultimately lead to atrial contraction.
- The QRS complex shows the largest deviation from the isoelectric line and represents the depolarization of the ventricles, which will ultimately lead to ventricular contraction. The QRS complex provides the largest spike in the ECG trace because of the large muscle mass that exists in the ventricles.
- The T-wave represents ventricular repolarization.
- Atrial repolarization is not commonly seen on ECG traces as it occurs during ventricular depolarization and takes place during the period of the QRS complex.
- In addition to the main electrical events depicted in the ECG, a number of time intervals can also provide insight into the functioning of the conduction system.
- The time from the P-wave to the R-wave (the P–R interval) is normally 0.15 seconds and represents the time taken for action potentials to be propagated into the ventricles. Therefore a substantially increased P–R interval may indicate a problem with the atrioventricular node.

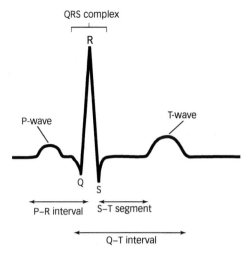

Figure 5.7 A typical ECG trace showing the main events and time intervals

- The time taken to complete the QRS complex represents the time it takes for action potentials to propagate throughout the ventricles. Increases in this time may indicate problems with the bundles of His and/or Purkinje fibres.

Looking for extra marks?

The ECG trace can be utilized to determine a variety of problems with the conduction system. Some of the more common conduction system disorders are sinus node dysfunction and atrioventricular block. Sinus node dysfunction indicates a problem with the sinoatrial node that leads to failure to generate or propagate action potentials. It may present as severe bradychardia (reduction in the number of P-waves), sinus node arrest (no P-waves present), exit block (P-wave present but unable to propagate to ventricles), or tachycardia–bradycardia syndrome (burst of P-wave activity followed by reductions in P-wave activity). Atrioventricular block occurs at the level of the atrioventricular node and is classified according to the extent of the problem. This may range from reduced ability to propagate action potentials into the ventricles to complete loss of synchronization between the contraction of the atria and ventricles. First degree atrioventricular block usually occurs at the level of the atrioventricular node and can be seen as an increase in the P–R interval on an ECG. This represents a reduced ability to propagate action potentials into the ventricles. Second degree atrioventricular block usually occurs at the level of the atrioventricular node and results in intermittent propagation of action potentials into the ventricles. Third degree, or complete, atrioventricular block represents a failure to propagate action potentials from the atrioventricular node, resulting in no synchronization of atrial and ventricular contraction. Bundle branch block is shown as an increase in the time taken to complete the QRS complex.

5.7 THE CARDIAC CYCLE

The cardiac cycle refers to the cyclical process from the start of one heart contraction to the start of the next. As outlined previously, the constant changes in pressure that occur throughout this process are key to ensuring that blood is moved through the system efficiently.

- At rest, one cardiac cycle takes approximately 0.7–0.8 seconds, although this depends on the individual in question.
- One cardiac cycle is split into systole (ejection phase) and diastole (filling phase). Systole accounts for approximately one-third of a cardiac cycle with diastole accounting for two-thirds.

Ventricular systole

- Prior to the start of ventricular systole, the atria and ventricles are relaxed with the atrioventricular valves open but other valves closed. The atrioventricular valves close when the pressure inside the ventricles exceeds the pressure in the atria. Immediately before the start of ventricular systole, blood is in the ventricles. The volume of this blood is referred to as the end-diastolic volume and usually amounts to approximately 120–130mL.
- Ventricular systole consists of two periods: a period of isovolumic contraction and a period of ejection.
- The period of isovolumic contraction follows the QRS complex on an ECG trace. During this period, there is an increase in pressure within the ventricles due to contraction of the myocardium. However, blood does not eject from the heart because the aortic and pulmonary valves are closed.
- When the pressure inside the ventricle exceeds the pressure within the arteries, the aortic and pulmonary valves open and this leads to the period of ejection.
- During the initial stages of the period of ejection, blood flow is very fast because of the pressure gradient between the ventricles and arteries. As this pressure gradient reduces, largely due to ventricular relaxation and consequent reduction in pressure within the ventricle, the rate of blood flow reduces until the pressure within the arteries exceeds the pressure in the ventricles. At this point, the aortic and pulmonary valves close. The volume of blood remaining in the ventricles at this point is referred to as the end-systolic volume and usually amounts to approximately 50–60mL.

Preload and afterload

- **Preload** (otherwise known as Starling's law of the heart) refers to the importance of end-diastolic volume in determining stroke volume.
- As end-diastolic volume increases, stroke volume increases. This is largely due to an increase in elasticity of cardiac muscle fibres prior to contraction, which results in a greater force of contraction and ultimately a greater stroke volume.

- End-diastolic volume is determined by venous return. Thus an increase in venous pressure leads to an increase in venous return, which leads to an increase in end-diastolic volume and ultimately to an increase in stroke volume.
- **Afterload** refers to the force against blood flow when blood is ejected from the left ventricle. This is determined primarily by aortic pressure. If aortic pressure is relatively high, the pressure gradient that exists between the left ventricle and the aorta is relatively low, which leads to reduced blood flow.
- While an increase in preload increases stroke volume, an increase in afterload reduces stroke volume.
- Exercise tends to increase blood pressure and thus leads to an increase in afterload. Despite this, stroke volume is increased during exercise because of an increase in venous return and the consequent increase in preload.

Ventricular diastole

- Ventricular diastole consists of two periods: a period of isovolumic relaxation and a period of filling.
- The period of isovolumic relaxation occurs after the pulmonary and aortic valves have closed. It follows the T-wave of an ECG and signifies a reduction in ventricular pressure without any change in blood flow.
- When ventricular pressure reduces to less than atrial pressure, the atrioventricular valves open and this leads to a period of filling.
- Initially, ventricular filling is passive and this accounts for about two-thirds of the filling of the ventricle. Following the P-wave on an ECG, the atrial myocardium contracts, causing blood to move into the ventricle until atrial pressure exceeds ventricular pressure. At this point, the atrioventricular valves close and ventricular systole begins again.

Check your understanding

Outline the difference in structure and function between arteries and veins. (*Hint: Consider pressures and direction of blood flow*)

Describe the main factors that influence mean arterial pressure. (*Hint: Consider the main site of resistance to blood flow and factors that affect cardiac output*)

How might blood pressure be influenced by lifestyle factors? (*Hint: Consider the effects of diet, physical activity, and stress*)

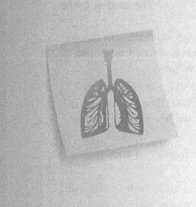

6 The respiratory system

The principal role of the respiratory system is to provide an exchange of gases between the body and the environment. It also has a number of other functions; two examples are a contribution to the maintenance of plasma pH and the production of sound. In exchanging gases, it ensures that adequate amounts of oxygen are delivered to tissues and, equally, that carbon dioxide is efficiently removed. It manages to do this when faced with a variety of environmental challenges (e.g. exercise).

Key concepts

- The paired lungs sit inside the thorax. They are formed from a series of bifurcations of a single trachea. Some regions of the lung are conductive, some are involved in gas exchange, and others are a mixture.
- Air enters the lung by a suction pump. Inspiration results in an increase in volume and therefore a decrease in pressure inside the lungs. Since air is at a higher pressure in the environment, it enters the lungs down its pressure gradient.
- Whilst inspiration is an active process, expiration, at least at rest, is a passive process which depends on the elasticity of the lungs and thorax.
- Gas exchange between the environment and plasma takes place in the alveoli. Again, pressure gradients move oxygen into and carbon dioxide out of the plasma.

- Gas solubility is relatively limited, so in order to increase the total amount of oxygen that can be carried in plasma, red blood cells contain the oxygen transport pigment haemoglobin. Oxygen binds to haemoglobin and this produces a significant increase in the amount of oxygen carried in plasma.
- The majority of carbon dioxide is transported in plasma in the form of the bicarbonate ion. Smaller amounts are linked to haemoglobin and even less is dissolved in plasma.
- Central and peripheral chemoreceptors maintain plasma oxygen and carbon dioxide at appropriate levels.

6.1 ORGANIZATION OF THE RESPIRATORY TRACT

The respiratory system, which consists of the paired lungs, enables gas exchange to occur between the blood plasma and the environment. It also has a number of other roles; for example, it contributes to the acid–base balance and phonation.

General organization

- The paired lungs are found in the **thoracic cavity**–the right lung has three lobes whilst the left lung has two lobes.
- Air enters the respiratory tract through the nose or mouth, and then via the **pharynx** and **larynx** to the **trachea**.
- The trachea is a tube-like structure about 15–18cm long and 1.5cm in diameter. There are incomplete rings of cartilage within its walls which prevent it from collapsing. The region which has no cartilage associated with it is adjacent to the oesophagus. The absence of cartilage here allows food to pass through the oesophagus unhindered.
- Cartilage is present up to and including the **bronchioles**, although in smaller bronchi and bronchioles plates of cartilage rather than rings provide rigidity.
- The trachea then undergoes as series of bifurcations. It splits into two main bronchi–one left and one right. Each of these then split into two and so on. In total there are 23 generations of airways (Figure 6.1).
- Smooth muscle is present in all generations up to the **alveolar ducts**.

Microscopic organization

- The epithelium of the respiratory tract down to the small bronchi consists of ciliated epithelium.
- The cilia are motile and work to trap inhaled particles that have evaded capture in the nose.
- Particles are trapped within the mucus secreted by goblet cells. The motility of the cilia moves the mucus up the airways to the throat.
- The microscopic structure of the ciliated region of the airways is shown in Figure 6.2.

Organization of the respiratory tract

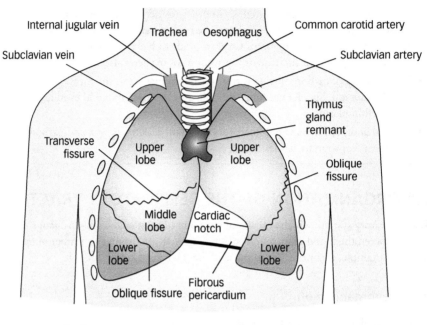

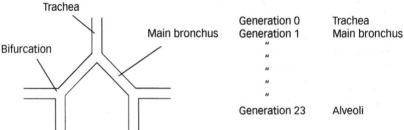

Figure 6.1 Overall organization of airway structure

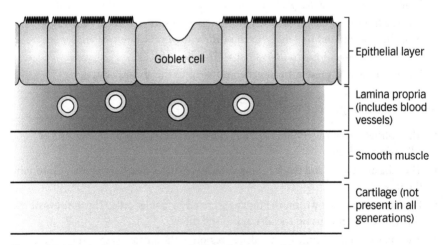

Figure 6.2 Histological appearance of ciliated airway epithelia

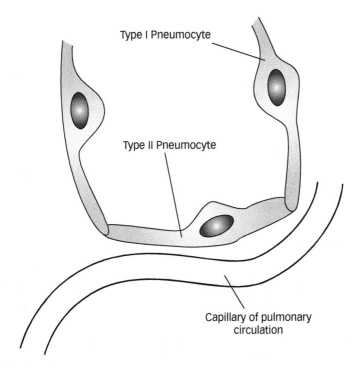

Type I Pneumocyte

Type II Pneumocyte

Capillary of pulmonary
circulation

Figure 6.3 Alveolar structure

- Gas exchange occurs in the most distal region of the airways–the **alveoli**.
- The total surface of the alveoli is about $70m^2$–this, together with a vast number of capillaries, ensures that gas exchange proceeds unhindered.
- The alveoli have the thickness of a single cell and are made up of cells called type I and type II **pneumocytes**.
- Type I pneumocytes are the basic building blocks of the alveoli.
- Type II pneumocytes secrete **surfactant**, which is a detergent-like substance. Its role is to reduce the surface tension that is generated at air–water interfaces, such as those that are present in the alveoli. In the absence of surfactant the alveoli would probably collapse, and this would inhibit gas exchange.
- The organization of the alveoli is shown in Figure 6.3.

Looking for extra marks?

The smooth muscle of the airways can be influenced by changes in the activity of the autonomic nervous system. There is a direct innervation by the parasympathetic division. When receptors of the sympathetic division are activated, the smooth muscle relaxes—bronchodilation. However, although there are receptors (β receptors) for the neurotransmitters of the sympathetic division, there are no nerves innervating them. Rather, they respond to changes in circulating plasma levels of adrenaline and noradrenaline. The smooth muscle contracts in response to activation of the parasympathetic nervous system—bronchoconstriction. This phenomenon is utilized in the treatment of asthma. Inhalation of drugs which activate β receptors (e.g. salbutamol) relax the constriction of airways, which occurs during an asthma attack, thus making breathing easier.

6.2 MECHANICS OF BREATHING

Gases move between the environment and blood plasma (and vice versa) down their partial pressure gradients. Air moving into the lungs is a purely passive process, but air movement out may be aided by muscular activity.

The lungs inside the thorax

- In order for gas exchange to occur, air must be drawn into the lungs and subsequently exhaled–this happens on a regular basis, about 12 times every minute.
- As will be seen, air is drawn into the lungs by changes in volume, and therefore pressure, within the thorax.
- In order for the volume of the thorax to change it is necessary for it to move, and in order to achieve that movement muscular contraction must occur.
- The principal muscles of respiration are the **diaphragm** and the **respiratory intercostal muscles**.
- The diaphragm is a large sheet of muscle, which separates the thorax from the abdomen. At the end of an expiration, the diaphragm has a dome-like appearance. During inspiration the diaphragm contracts and flattens. The diaphragm contracts in response to activity in the phrenic nerve.
- The respiratory intercostal muscles are located in the spaces between the ribs. There are two types–the inspiratory intercostals and the expiratory intercostals. The inspiratory intercostals lift the ribs upwards and outwards. The expiratory intercostals move the ribs in the opposite direction.
- In order to facilitate expansion of the lungs, both the lungs and the inside of the thorax are lined with a specialized type of membrane called **pleura**.
- The lungs are lined with visceral pleura whilst the inside of the thorax is lined with parietal pleura.
- Between the pleura is a fluid-filled space called the pleural cavity. The role of the fluid is to allow the lungs to move within the thorax during inspiration and expiration.
- However, whilst the pleura allow the lungs to move easily and freely within the thorax, they also essentially attach them to the inside of the thorax. A good example of this is to take two microscope slides. If a drop of water is placed on one and the other slide is placed on top of it, the slides move easily over each other, but it is difficult to prise them apart.
- The organization of the lungs within the thorax is shown in Figure 6.4

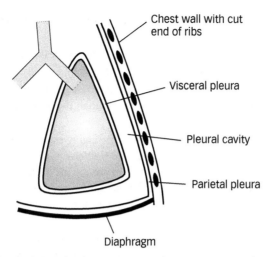

Figure 6.4 The lungs within the thorax showing the organization of the pleural membranes

Volume and pressure changes associated with breathing

- Ventilation is a cyclical process consisting of alternate inspirations and expirations—air moves into the lungs during inspiration and out during expiration.
- Air moves because of differences in pressure—it moves from regions of high pressure to regions of lower pressure. When we talk about high and low pressure, we are referring to changes in pressure above and below normal atmospheric pressure, respectively. Normal atmospheric pressure is taken to be 760mmHg.
- Pressure is a measure of the force of a gas. This is most easily understood by thinking about a closed container containing air. The air that we breathe is a mixture of nitrogen, oxygen, carbon dioxide, and various other gases. The pressure of air within that container is simply a measure of the number of collisions between gas molecules in the air and the sides of the container.
- If the volume of the container is decreased, but the volume of air is the same, the pressure of the gas increases—the molecules in the air collide with the sides of the container more frequently. If the volume of the container increases, the opposite happens.
- Therefore the pressure of a gas is inversely proportional to its volume—this is known as Boyle's law.
- At the beginning of every respiratory cycle, i.e. at the beginning of inhalation, the pressure inside the lungs is equal to the atmospheric pressure. Since there is no difference in pressure, there is no movement of air.
- As inspiration proceeds the diaphragm contracts and flattens, and the inspiratory intercostal muscles move the ribs up and out. The net effect of this is that the volume of the thorax increases.

Lung volumes

- Because of the pleural linings, as the volume of the thorax increases, the lungs follow, and therefore lung volume increases.
- Because of Boyle's law, the pressure inside the lungs decreases, i.e. it falls below atmospheric pressure (it becomes sub-atmospheric).
- Therefore there is a pressure difference—the pressure in the mouth and nose is equal to atmospheric pressure, which is greater than the pressure inside the lungs. Consequently air moves into the lungs.
- At some point, inspiration stops. The lungs are highly elastic structures and now begin to return to their original volume. The elasticity of the lungs is such that they would collapse almost entirely if it were not for the presence of the pleural fluid holding them to the walls of the thorax. Sometimes (e.g. as a consequence of trauma) the pleural cavity may become breached and as a consequence the lungs collapse. This is called a pneumothorax.
- As a consequence of the lungs collapsing and returning to their original size, the pressure of the air inside them increases, i.e. it rises above atmospheric pressure. Now there is an outwardly acting pressure gradient and air is forced out of the lungs
- Thus inspiration is an active process requiring the coordinated contractions of muscles. However, expiration is a purely passive process caused by the inherent elasticity of the lungs.

Looking for extra marks?

The ease with which the lungs expand and contract during normal breathing is called their compliance. The greater the compliance, the less force is required to initiate inspiration and expiration. Compliance may change during a variety of disease states. For example, some diseases reduce surfactant production. This means the alveoli collapse, the lungs become stiffer, and greater pressure changes are required to inflate them.

The pressure within the lungs is called the intrapulmonary pressure. This is the pressure which decreases and increases during normal breathing. The changes in pressure are not very large. However, they may change significantly during various manoeuvres (e.g. the Valsalva manoeuvre). This is where a forced expiration occurs against a closed glottis, which prevents air from leaving the lungs. This causes a large rise in intrapulmonary pressure.

6.3 LUNG VOLUMES

Lung volumes, which vary depending on a number of factors (e.g. age, gender, etc.), can be measured using **spirometry**. Spirometry is also a useful clinical test for initial investigation of lung disease.

- Normal quiet breathing is called eupnoea. However, it is apparent from everyday circumstances that the volume of air inhaled and exhaled may vary considerably; for example, consider the changes in the pattern of breathing during exercise.

- Lung volumes can be measured with a spirometer. A number of 'different' lung volumes can be described. A typical spirometer trace is shown in Figure 6.5
- The **tidal volume (VT)** is the volume of air that is inspired and expired at rest—on average it is about 500mL.
- The **inspiratory reserve volume (IRV)** is the additional volume of air that can be inspired over and above that of the normal tidal volume—this may 3000–3500mL.
- The **expiratory reserve volume (ERV)** is the volume of air that can exhaled over and above a normal tidal volume—this may be about 1000mL.
- The **vital capacity (VC)** is the sum of VT + IRV + ERV and is about 5000mL.
- The **residual volume (RV)** is the volume of air remaining in the lungs after a maximal expiration—about 1000mL. This means the lungs can never be fully be emptied. The fact that this air remains in the lungs means that gas exchange is a continuous rather than a cyclical process. Its presence also helps to prevent alveoli from collapsing.

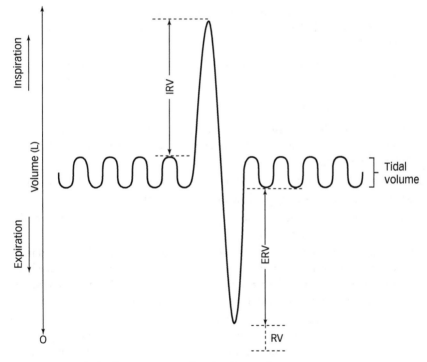

Figure 6.5 A typical spirometer trace showing lung volumes

Looking for extra marks?

Spirometry can be performed in a variety of ways—one of these measures dynamic lung volumes. Essentially, an individual takes in the biggest breath they can and then empties the lungs as fast and as forcefully as they can. The volume that is exhaled is called the forced vital capacity (FVC). Individuals in normal health are able to exhale 80–90% of their FVC in the first second of this manoeuvre. In individuals with asthma this value is significantly reduced—it may drop to 50–60%. Asthma is an inflammatory disease of the airways—the diameter of the airway reduces due to swelling, leakage of plasma, and production of increased volumes of mucus. This makes it difficult for air to move into and out of the lungs when measured by spirometry. The resistance to airflow is said to have increased.

6.4 THE COMPOSITION OF AIR AND GAS EXCHANGE

Composition of air

- The air that we breathe is a mixture of gases—it is made up of nitrogen (79%) and oxygen (20.9%), with the remainder being carbon dioxide, water vapour, and other gases (e.g. neon).
- We have previously stated that atmospheric pressure when measured is 760mmHg. Since air is a mixture of gases, each gas makes a contribution to atmospheric pressure. Each gas does so in relation to its percentage composition of the total gas. This is known as Dalton's law of partial pressures.
- Given that atmospheric pressure is 760mmHg and that oxygen contributes 20.9% to the composition of air, the partial pressure of oxygen in air is $20.9/100 \times 760 = 159$mmHg.
- The partial pressures of all other gases in air can be calculated in a similar manner.
- Partial pressure is denoted by p. Therefore $p\text{O}_2 = 159$mmHg.
- However, although humans breathe air, the gases that are inhaled are transferred to a liquid medium (i.e. plasma).
- The partial pressure of a gas will influence the amount of gas which dissolves in a fluid—this is known as Henry's law. In fizzy drinks, carbon dioxide is added to the drink under pressure. When a can or bottle containing this drink is opened it fizzes—the pressure is released and the gas comes out of solution.
- Another important factor influencing the gas content of liquids is the temperature of the fluid. As water is heated, bubbles start to appear—this is dissolved gas coming out of solution. In humans, this is not significant as body temperature remains at 37°C.
- Finally, the solubility of the gas also helps to determine how much dissolves in solution—the greater the solubility, the greater the amount of gas which dissolves.

- So, whilst partial pressures are important in determining the movement of gases (remember that gases will move from higher to lower partial pressures), solubility and temperature will also determine the actual volume of gas which dissolves in solution

Gas exchange and transport—oxygen

- As indicated previously, gas will move from regions of higher partial pressure to regions of lower partial pressure—this occurs by simple diffusion, which is a purely passive process
- As air enters the lungs, it will encounter the air that has remained in the lung from the previous expiration. There is mixing of this 'new' and 'old' air. Therefore the composition of air that reaches the alveoli is different from that originally inspired.
- The po_2 in the alveoli is about 100mmHg—blood flowing through the lungs has a lower po_2 of about 40mmHg. Therefore oxygen will move from the alveoli to the capillaries of the pulmonary circulation. Blood from here will also mix with blood from other regions of the pulmonary circulation where there is no exchange. This means that blood entering the left side of the heart has a po_2 of about 90mmHg.
- As blood enters tissues, it encounters interstitial fluid with a po_2 of about 40mmHg. Therefore oxygen now moves from plasma into interstitial fluid and thence into cells.
- Despite sufficient partial pressures moving oxygen from air to plasma to cells, oxygen has only a limited solubility in plasma. The amount that dissolves would be insufficient to service the body, given its oxygen requirements. Therefore, in order to increase the amount that can be carried in blood, red blood cells contain haemoglobin.
- Haemoglobin (Hb) is an oxygen transport pigment. It is a conjugate protein, i.e. a protein which has prosthetic groups associated with it.
- The protein component of Hb is globin and the prosthetic part is haem. The Hb molecule is a tetramer—it consists of four globin molecules, each of which is associated with a haem group.
- Haem is a porphyrin-type molecule which has at its centre an iron ion in the Fe^{2+} state.
- The Hb content of plasma is about 14g/100mL, which means that the oxygen-carrying capacity of blood is about 20mL O_2/mL blood. Without Hb it would only be a fraction of this. A typical red blood cell contains about 300 million Hb molecules.
- Oxygen combines reversibly with the haem component of Hb. Each haem group combines with one O_2 molecule, so in total each Hb molecule combines with four O_2 molecules.
- Experimentally, it is possible to take a sample of blood and measure the proportion of Hb molecules bound to it at varying po_2 values—this is called an **oxygen–haemoglobin** saturation curve (Figure 6.6).

The composition of air and gas exchange

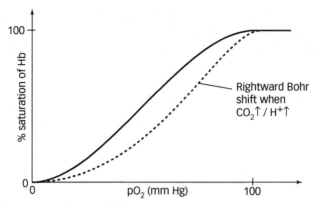

Figure 6.6 The oxygen–haemoglobin saturation curve

- The O_2–Hb saturation curve is an S-shaped or sigmoidal curve. This is because Hb displays cooperativity. The binding of the first molecule of O_2 produces a conformational change which makes the binding of the second easier and so on until all four molecules are bound.
- As can be seen from Figure 6.6, Hb is almost fully saturated at a po_2 of about 90–100mmHg–the po_2 in the alveoli. Haemoglobin which is fully saturated with oxygen is called oxyhaemoglobin, whilst that with no oxygen is called deoxyhaemoglobin.
- As indicated previously, the po_2 of tissues is about 40mmHg. At this partial pressure, Hb is still about 70–80% saturated with oxygen. This means that there is a substantial 'reserve' of oxygen available if needs change (e.g. during exercise).
- As suggested earlier, a number of factors can influence the degree of saturation of Hb. One of these is a drop in pH. As pH decreases, the O_2–Hb dissociation curve moves to the right—for a given po_2, there is a reduction in saturation, i.e. more O_2 has been released. This can be seen in Figure 6.6. This rightward shift in the O_2–Hb dissociation curve is called the **Bohr effect**. The effect of increased levels of CO_2 produces the same effect. This is because CO_2 is an acidic gas, i.e. $CO_2 + H_2O \leftrightarrow H_2CO_3 \leftrightarrow H^+ + HCO_3^-$. An increase in temperature also produces a Bohr effect.
- This drop in pH/increase in CO_2 is also seen in, for example, exercising tissues. The significance of this is that active tissue Hb gives up more of its oxygen, thus ensuring that adequate oxygen is delivered to the tissues where it is needed.
- Red blood cells contain a substance called **2,3-DPG** (2,3-diphosphoglycerate). This is a product of metabolism within the mitochondria of red blood cells. A basal level of this compound is present at all times, and it has a direct effect on O_2 binding to Hb. When its levels rise (e.g. due to a variety of endocrine stimuli or a reduction in atmospheric oxygen content) the release of O_2 from Hb is encouraged.

> ## *Looking for extra marks?*
>
> The globin element of Hb exists in different forms. In adults, a molecule of Hb consists of two α-globin chains and two β-globin chains. In contrast, fetal Hb comprises of two α and two γ chains. Fetal Hb has a much higher affinity for O_2 than adult Hb: its O_2–Hb dissociation curve lies to the left of that of the adult curve.
>
> Whilst O_2 combines reversibly with Hb, carbon monoxide (CO) combines irreversibly with it. Furthermore, CO has a much higher affinity for Hb than O_2 does. Therefore exposure to CO at even relatively low concentrations effectively renders Hb ineffective at carrying O_2. This obviously compromises O_2 delivery to tissues and can prove to be fatal. Treatment to overcome this includes the delivery of hyperbaric oxygen.

Gas exchange and transport—carbon dioxide

- Carbon dioxide is a waste product of cellular metabolism. Whereas the respiratory system is responsible for delivering O_2 into the plasma, its responsibility in relation to CO_2 is to excrete it from the plasma to the environment.
- Unlike O_2, CO_2 can be transported back to the lungs by a variety of mechanisms prior to its exhalation. However, like O_2, CO_2 moves down its concentration gradients.
- The majority of CO_2 is transported back to the lungs in the form of **bicarbonate ions (HCO_3^-)**. Carbon dioxide passes from tissue cells into red blood cells. Here, under the influence of the enzyme carbonic anhydrase, it forms carbonic acid, which in turn dissociates into hydrogen ions and bicarbonate ions: $CO_2 + H_2O \leftrightarrow H_2CO_3 \leftrightarrow H^+ + HCO_3^-$ The hydrogen ions bind to Hb, whilst the HCO_3^- is transported into the plasma in return for an influx of Cl^- ions. This is called the **chloride shift**. In the lungs, all reactions are reversed and CO_2 is exhaled.
- About 75% of CO_2 transport occurs via the bicarbonate ion.
- A second, relatively large, means by which CO_2 is transported is by CO_2 binding directly to Hb to form a compound called **carbaminohaemoglobin**. The CO_2 binds to free amino groups within the globin units of the Hb molecule. This mechanism accounts for just over 20% of CO_2 transport.
- The final way in which CO_2 is transported back to the lungs is in simple solution, i.e. CO_2 dissolved in plasma. However, there is limited capacity for this, and less than 10% of CO_2 is transported by this method.
- A summary of CO_2 transport is presented in Figure 6.7.

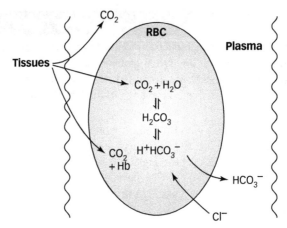

Figure 6.7 Summary of carbon dioxide transport mechanisms

6.5 CONTROL OF BREATHING

It is apparent from everyday life that breathing is a cyclical process and that the rate at which it occurs can be varied to match a range of requirements (e.g. increased O_2 demands during exercise).

Generation of respiratory rhythm

- The normal respiratory rate is between 12 and 15 breaths/minute. Hypo- and hyperventilation are rates below and above this value, respectively.
- The respiratory rhythm is generated by neurons located in the lower part of the brainstem, i.e. in the medulla. Two groups of neurons are important—the dorsal respiratory group and the ventral respiratory group. Together they constitute the **respiratory centre**.

- The **dorsal respiratory group** generate action potentials. These neurons fire during the inspiratory phase of breathing and therefore are considered to be inspiratory neurons. The neurons synapse on neurons of the phrenic nerve—action potentials in this nerve produce contraction of the diaphragm.

- The **ventral respiratory group** are both inspiratory and expiratory. They receive an input from the dorsal group and themselves synapse on neurons which form the phrenic nerve and also the intercostal nerves. Activity here is important when additional demands are placed on breathing (e.g. during exercise). Equally, it is under conditions of this type that the activity of the expiratory neurons is important.

- Previously, we have stated that expiration is a passive process which depends on the inherent elasticity of the lungs and thorax. However, under certain circumstances (e.g. exercise) expiration may be active, i.e. neutrally driven muscular activity helps in the exhalation process. For example, there are expiratory intercostal muscles. During extreme demands other muscles also play a role (e.g. muscles of the abdomen).

- Given the role that the dorsal respiratory group plays in inspiration, it might seem reasonable to conclude that this is the origin of the drive to breathe. However, it is now thought that the ventral respiratory group plays this role–in particular a region called the **pre-Botzinger complex.**

- The mechanism by which the pre-Botzinger complex initiates the respiratory rhythm is unclear.

- There are regions elsewhere in the brainstem that modulate the basic pattern of breathing.

- There is a region in the pons called the **pneumotaxic centre**. Neurons from here synapse on neurons in the dorsal respiratory group and their role is to inhibit activity in inspiratory neurons. Likewise, activity in the **apneustic centre** prevents inhibition of inspiratory neurons, therefore prolonging inspiration. In normal breathing, the role of the rate-limiting pneumotaxic centre is most important.

- The basic organization of the centres involved in the generation of breathing is shown in Figure 6.8.

Control of the respiratory rate

- Given that the role of the respiratory system is to maintain appropriate levels of CO_2 and O_2 in the plasma, it is entirely appropriate that these compounds influence the pattern of breathing.

- 'Chemoreceptor' is the general term given to sensory structures which monitor levels of chemicals, such as O_2 and CO_2, in the body. There are separate receptors for each of these two chemicals and changes in either can influence the pattern of breathing.

- Oxygen levels are monitored by chemoreceptors located in the carotid body and the arch of the aorta.

Control of breathing

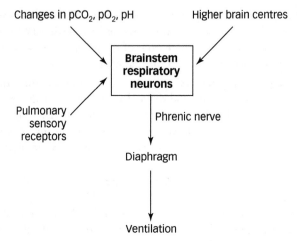

Figure 6.8 Location and organization of peripheral chemoreceptors

- The **carotid bodies** are found where the common carotid artery divides to form an internal and an external carotid artery. Afferent neurons from the carotid bodies travel in the 9th cranial nerve, the glossopharyngeal nerve. These neurons terminate in the respiratory centres of the medulla (Figure 6.9).
- The carotid bodies increase afferent discharge when there is an increase in plasma po_2 or pco_2, or a decrease in pH. Although they are responsive to changes in both CO_2 and O_2, it is their role in monitoring O_2 levels which is most important.
- When plasma po_2 levels decrease, the carotid bodies increase their discharge rate. This in turn stimulates the respiratory centre in the medulla to increase both the rate and depth of breathing in an attempt to restore plasma po_2 to its desired level.

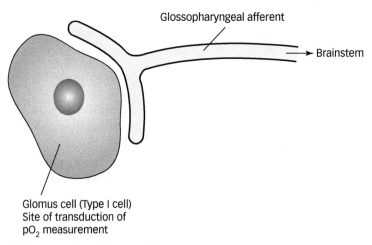

Figure 6.9 Organization of the carotid body

- Plasma po_2 level must fall by a significant amount—to about 60mmHg—before the carotid bodies are activated. This represents a drop of about 40% in plasma po_2. A further drop to about 40mmHg will only increase respiratory rate by about 50%. This indicates that the carotid bodies are relatively insensitive structures.

- The second group of chemoreceptors which monitor oxygen levels are the **aortic bodies**. Afferent neurons travel from these back to respiratory centres in the medulla in the 10th cranial nerve, the vagus nerve. They respond to the same stimuli and in the same way as the carotid bodies. However, there is virtually no response to reduced plasma po_2 in humans who have had their aortic bodies removed, and this raises questions about the role of the aortic bodies.

- Together, the carotid and aortic bodies are known as peripheral chemoreceptors.

- In contrast, the chemoreceptors which monitor plasma pco_2 levels are known as **central chemoreceptors**. This is because they are located on the surface of the medulla. Unlike the peripheral chemoreceptors, the central chemoreceptors only respond to changes in CO_2.

- As plasma pco_2 rises, it passes across the blood–brain barrier and enters the cerebrospinal fluid. Once in the CSF, it dissolves and ultimately forms bicarbonate ions and hydrogen ions, as described previously. The CSF has relatively poor buffering capabilities, and in fact it is the increase in hydrogen ion concentration which stimulates the central chemoreceptors. In turn, these activate neurons in the medullary respiratory centre to increase ventilation.

Looking for extra marks?

Much work has been done to investigate how the carotid bodies work. It is now known that the carotid bodies consist of two types of cell—type I and type II (Figure 6.8). Type II cells probably have a glial-like activity. The functional cells are the type I cells, sometimes called glomus cells. Type I cells are the transducers of the carotid bodies. Embedded in their membranes are O_2-sensitive K^+ channels. When po_2 drops, these channels begin to close and this causes their depolarization. This in turn results in Ca^{2+} influx and the release of neurotransmitters on to the afferent neurons. Release of a variety of neurotransmitters, including ATP, has been suggested.

Other factors influencing respiratory rate

- Whilst the principal factors influencing ventilatory rate are O_2 and CO_2 levels, there are a range of other stimuli which also influence the rate of ventilation.

- Within the respiratory tract, there are a number of different types of sensory receptor—stretch receptors, irritant receptors, and C-fibre receptors.

- Stretch receptors are not thought to play any significant role in controlling ventilation in humans. However, in smaller mammals these receptors, which are located in airway smooth muscle, are activated during inspiration. They send afferent signals back to the medullary respiratory neurons to inhibit ventilation

Control of breathing

- Irritant receptors are thought to be responsible for sneezing and coughing. Both of these responses require an increased inspiration followed by an expiration against a closed glottis. As intrapulmonary pressure increases, the glottis opens followed by high-velocity expiration.
- C-fibre receptors are so called because the afferent neurons associated with them are non-myelinated. They become activated during certain pathological states (e.g. pulmonary oedema and pulmonary embolism). Activation of these receptors results in rapid shallow breathing.
- Alongside the many stimuli which reflexly influence ventilation, it is also possible to influence ventilation voluntarily. Both the rate and depth of ventilation can be increased or decreased at will. This is of importance in speech production for example.

Check your understanding

Why are β_2 agonists effective in treating asthma? (*Hint: consider the structure of the airways and their innervation*)

Why can't residual volume be measured by spirometry? (*Hint: think about what RV is and the consequences of being able to empty the lungs fully*)

Identify the principal factors that control ventilation. (*Hint: think about the role of the lungs in maintaining homeostasis*)

7 Renal physiology

The most obvious role of the renal system is the production and elimination of urine. However, it plays a number of other vital roles in the human body. An understanding of the anatomy of the renal system is key to understanding its role in the body.

Key concepts

- The production and elimination of urine is the principle function of the renal system.
- The nephron is the functional unit of the kidney and is responsible for the formation of urine.
- The blood plasma is filtered within the nephrons and substances are secreted and reabsorbed in the process of urine formation.
- The extent of water and sodium loss via urine is controlled by the actions of arginine vasopressin, aldosterone, and atrial natriuretic peptide.
- The renal system plays a major role in the regulation of acid–base balance.
- An increase in blood pH leads to an increase in the quantity of bicarbonate secreted in urine in order to return blood pH to normal.
- A decrease in blood pH leads to an reduction in the quantity of bicarbonate secreted in urine and an increase in the amount of newly synthesized bicarbonate in order to return blood pH to normal.

7.1 RENAL ANATOMY AND PHYSIOLOGY

The principle function of the renal system is the production of **urine**. However, it has other important functions as well. An understanding of the anatomy of the renal system is important in order to understand its function.

- Urine consists of water, metabolic by-products, other chemicals, and ions. Consequently, the renal system is heavily involved in regulation of water and electrolyte balance.
- In addition, the renal system plays a major role in the excretion of unwanted metabolic by-products and acid–bases, such as urea and creatinine.
- The renal system plays a role in the control of acid–base balance via the absorption or secretion of hydrogen ions and bicarbonate.
- The renal system secretes a variety of hormones and, as such, is involved in regulation and control of various physiological functions.
- In certain situations glucose can be synthesized from amino acids in the **kidneys**, resulting in an increase in blood glucose concentration.

Anatomy of the kidneys

- The renal system consists of the kidneys (Figure 7.1), the **ureters**, the **bladder**, and the **urethra**.
- The kidneys are located in the dorsal abdomen between the T12 and L3 levels of the spine. The right kidney is located near the liver. Each kidney has an adrenal gland located on its upper surface.
- Each kidney is surrounded by an inner capsule of collagen fibres, a layer of adipose tissue, and an outer layer of collagen fibres that attach to surrounding tissue. This provides stability for each of the kidneys.
- The outer portion of each kidney is known as the **renal cortex** and the inner portion is known as the **renal medulla**.
- The renal medulla contains numerous **renal pyramids**, which are separated from each other by **renal columns**.
- Urine is produced and moved to the tip of each renal pyramid before being transported into a **minor calyx**. A **major calyx** is formed from numerous minor calices before entering the renal pelvis. The **renal pelvis** then drains into the ureter.

The nephron

- The **nephron** is the basic functional unit of the kidney. There are approximately 1 million nephrons per kidney.
- Each nephron consists of a **renal corpuscle** and a **renal tubule** (Figure 7.2). The purpose of the renal corpuscle is to filter the blood. This filtrate then enters the renal tubule where various substances are added to it, producing urine. The renal

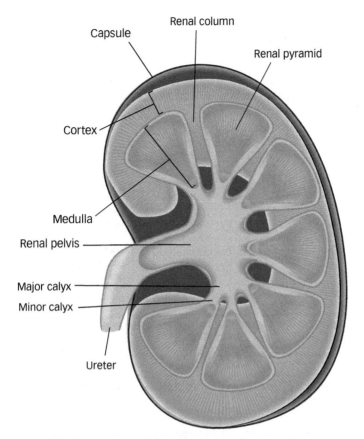

Figure 7.1 Schematic representation of the kidney showing the main structures.

tubule drains into a collecting duct which enters a minor calyx and ultimately leads to movement of urine into the ureter.

- There are two types of nephron in the kidneys: juxtamedullary nephrons and cortical nephrons.
- A juxtamedullary nephron is relatively long and its renal corpuscle is found close to the medulla of the kidney. Consequently, the renal tubule extends far into the medulla.
- A cortical nephron is relatively short and its renal corpuscle is located close to the cortex, or exterior, of the kidney. Its renal tubule does not extend far into the medulla. Approximately 85% of nephrons are cortical nephrons.
- A renal corpuscle has two sections: **Bowman's capsule** and the **glomerulus.**
- The glomerulus is essentially a bundle of capillaries. Blood enters a glomerulus via an afferent arteriole. While in the glomerulus, blood is filtered before it leaves via an efferent arteriole.
- Bowman's capsule is a storage site for the filtrate from the glomerulus. This filtrate, which is termed **glomerular filtrate**, moves from the glomerulus into

Renal anatomy and physiology

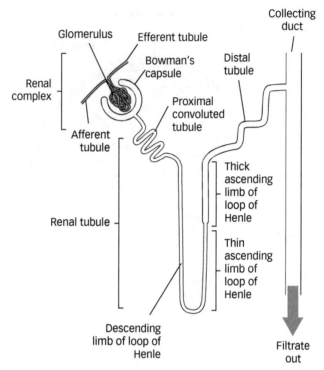

Figure 7.2 Schematic representation of a nephron showing sections of the renal corpuscle and renal tubule.

Bowman's capsule before entering the renal tubule. Bowman's capsule consists of an outer parietal layer of cells and an inner visceral layer of cells. The visceral layer acts as a filtration mechanism and consists of three cell layers. A single layer of endothelium lines the capillary before a basal lamina layer. The final layer consists of specialized cells called podocytes.

- The renal tubule has four distinct sections: the **proximal tubule**, the **loop of Henle**, the **distal tubule**, and the **collecting duct** system.
- The proximal tubule consists of two distinct sections—the proximal convoluted tubule and the proximal straight tubule.
- The loop of Henle consists of three distinct sections—the descending limb, the thin ascending limb, and the thick ascending limb.
- The distal tubule continues from the thick ascending limb of the loop of Henle. At the point where the ascending limb becomes the distal tubule, specialized cells form an area called the macula densa.
- The distal tubule is very close to the renal corpuscle and passes between the afferent and efferent arterioles that supply the glomerulus. A specialized layer of smooth muscle cells, called juxtaglomerular cells, can be found at the point where the afferent arteriole enters the renal corpuscle.

- The juxtaglomerular cells on the afferent arteriole and the macula densa on the distal tubule form the **juxtaglomerular apparatus.**
- The collecting duct system has two distinct sections—the cortical collecting duct and the medullary collecting duct. The difference between these sections reflects the positioning of the duct within the kidney.
- The kidneys have an unusual blood supply. As stated previously, blood enters the glomerulus via an afferent arteriole and leaves it via an efferent arteriole. This efferent arteriole then gives rise to the **peritubular capillaries.** The peritubular capillaries surround the proximal and distal tubules. The vasa recta is a specialized section of the peritubular capillaries that extend along the length of the loop of Henle. From the peritubular capillaries, blood empties into the interlobular veins, which ultimately lead to the renal vein and then the inferior vena cava.

Anatomy of the ureters, bladder, and the urethra

- A ureter extends from each kidney to the bladder. The ureter is a muscular tube which acts as a transport mechanism for urine. Peristaltic contractions occur, which assist in the movement of the urine into the bladder.
- The bladder acts as a storage facility for urine. As in the stomach, the bladder has numerous folds when it is not filled, which allow distension to occur.
- The ureters enter the bladder at the ureteral openings, which form a triangle with the internal urethral sphincter. This sphincter guards the entrance to the urethra, which is located at the base of the bladder.
- In males, there are three sections of the urethra: the prostatic, membranous, and spongy urethra. These sections are found as the urethra extends through the prostate gland, the urogenital diaphragm, and the penis, respectively. The end of the urethra is called the external urethral orifice.
- In females, the urethra is much shorter. It passes through the urogenital diaphragm and the external urethral orifice is located anterior to the vagina.
- The external urethral sphincter is one of the only sphincters in the body to be under voluntary control. In both sexes, it is found where the urethra passes through the urogenital diaphragm.

7.2 URINE FORMATION

As the main function of the renal system is the production and excretion of urine, it is important to understand how the blood is filtered.

Glomerular filtration

- The process of urine formation begins in the glomerulus with the filtration of blood plasma to produce the glomerular filtrate.

Urine formation

- The amount of glomerular filtrate produced depends on the percentage of cardiac output that is being delivered to the kidneys. This percentage is also known as the **renal fraction** and tends to be between 10% and 30%.

- The percentage of plasma that is filtered by the glomerulus, the filtration fraction, is approximately 19–20% resulting in a glomerular filtration rate (GFR) of approximately 125mL/min or 180L/day. Since urine production is not this high, it follows that a significant amount of the glomerular filtrate is reabsorbed within the nephron.

- GFR is determined by a pressure gradient, the filtration pressure, that is present between the glomerulus and Bowman's capsule. In order for the plasma to cross the filtration barrier, the pressure inside the glomerulus must exceed the pressure within Bowman's capsule.

- The three pressures that determine the overall filtration pressure are the glomerular pressure, Bowman's capsule pressure, and the osmotic pressure resulting from unfiltered proteins present in the glomerulus.

- The glomerular pressure is usually relatively high whereas the pressure within Bowman's capsule is relatively low, which means that the filtration pressure almost always favours filtration of plasma into Bowman's capsule (Figure 7.3).

- GFR changes in response to contraction or relaxation of the smooth muscle around afferent and efferent arterioles supplying the glomerulus. This is achieved via hormonal or central nervous system innervation.

- Contraction of afferent arteriole smooth muscle leads to a reduction of blood flow into the glomerulus and a reduction in glomerular pressure. Consequently, filtration pressure is reduced and GFR is reduced.

- Contraction of efferent arteriole smooth muscle leads to pooling of blood within the glomerulus and an increase in glomerular pressure. Consequently, filtration pressure is increased and GFR is increased.

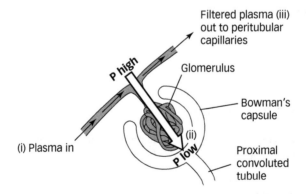

Figure 7.3 Filtration in the renal corpuscle: (i) unfiltered plasma enters the corpuscle via the afferent arteriole; (ii) plasma is filtered due to the pressure difference between the glomerulus and Bowman's capsule; (iii) filtered plasma leaves via the efferent arteriole and enters the peritubular capillaries where reabsorbed solutes, electrolytes, and water are added to the plasma

Tubular reabsorption

- GFR is the volume of filtrate produced that enters the renal tubule. As stated previously, GFR is approximately 180L/day, whereas urine production tends to be 1.5L/day. This suggests that approximately 99% of water within the glomerular filtrate is reabsorbed prior to the excretion of urine. This reabsorption takes place in the renal tubule. In addition to water, a variety of other solutes that are found in the glomerular filtrate are also reabsorbed in the renal tubule. This process is termed **tubular reabsorption**.

- Solutes that are reabsorbed in the renal tubule include, but are not limited to, amino acids, glucose, fructose, potassium, sodium, chloride, magnesium, and urea (Figure 7.4).

- Once reabsorbed, these substances enter the interstitial fluid before being moved into the peritubular capillaries that surround the renal tubules.

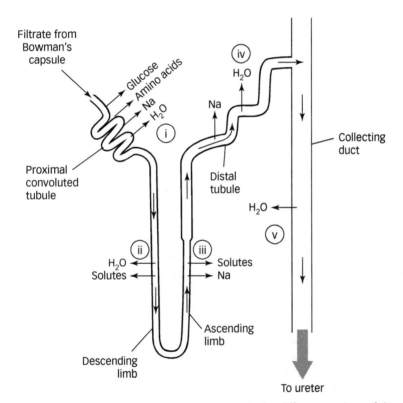

Figure 7.4 Solute, electrolyte, and water movement in the different sections of the nephron: (i) water, amino acids, sodium, and other solutes are reabsorbed in the proximal convoluted tubule; (ii) water and solutes are reabsorbed in the descending limb of the loop of Henle; (iii) sodium and solutes are reabsorbed in the ascending limb of the loop of Henle; (iv) sodium and water are reabsorbed in the distal tubule, depending on hormone concentrations; (v) water is reabsorbed in the collecting ducts depending on hormone concentrations

Urine formation

- The reabsorption of a number of these solutes, as well as the reabsorption of water, is under physiological control, whereas the reabsorption of others, such as glucose, is not. In these cases almost 100% of the substrate found in the glomerular filtrate is reabsorbed in the renal tubule and none is found in the urine. Therefore evidence of these substrates in the urine may indicate a pathology. For example, the presence of glucose in the urine is a sign of diabetes mellitus.

- The reabsorption of solutes occurs via diffusion, co-transport with sodium, counter-transport, or active transport. Water is reabsorbed via osmosis concurrently with the transport of solutes.

- The majority of reabsorption of solutes and water occurs within the proximal tubule.

- The vast majority of reabsorption of solutes within the proximal tubule occurs via co-transport with sodium. The concentration of sodium within the nephron cell is very low because of the active transport of sodium across the basal membrane in exchange for potassium. The concentration of sodium within the glomerular filtrate is very high, which results in a large concentration gradient between the lumen and the cell. The apical membrane of the nephron contains transporters for a number of solutes. The solutes can be reabsorbed via these transporters as a result of co-transport with sodium. The solutes are then diffused across the basal membrane into the interstitial fluid.

- The proximal tubule is permeable to water, so the movement of these solutes leads to reabsorption of water via osmosis.

- In the descending limb of the loop of Henle, the majority of reabsorption from the filtrate occurs as a result of diffusion. The concentration of solutes within the interstitial fluid in the medulla of the kidney is very high, which establishes a concentration gradient favouring diffusion of a number of solutes across the permeable wall of the descending limb. Again, water is reabsorbed by osmosis.

- Reabsorption of some ions is possible in the ascending limb via co-transport with sodium across the apical membrane and diffusion across the basal membrane. Water is not reabsorbed in this section of the nephron.

- Reabsorption of some ions also takes place in the distal tubule and the collecting ducts via active transport with sodium across the apical membrane and diffusion across the basal membrane. The permeability of the distal tubule and collecting ducts to water depends on the circulating concentrations of hormones, which will be covered later in this chapter.

- While some solutes, such as glucose, are completely reabsorbed from the glomerular filtrate, others such as urea and creatinine, are not. This ensures that the plasma level of these solutes does not increase to dangerous levels. Ultimately, they are excreted in the urine.

Tubular secretion

- In addition to the movement of solutes from the glomerular filtrate into the interstitial fluid, the renal tubule is also responsible for the secretion of substances from the peritubular capillaries into the glomerular filtrate.

- The majority of **tubular secretion** occurs in the proximal tubule.
- The principal substances secreted are hydrogen ions and potassium. Other substances such as ammonia and creatinine are also secreted.
- Substances are secreted via either diffusion or active transport.

7.3 WHOLE-BODY WATER AND SODIUM BALANCE

The renal system plays a major role in the control of whole-body fluid and sodium balance. This is achieved in a number of ways.

- Body water accounts for approximately 70% of an individual's body mass. Therefore a 70kg male consists of approximately 49L of water.
- One-third of total body water is found in the extracellular fluid with the remaining two-thirds in the intracellular fluid.
- Approximately a quarter of the water in the extracellular fluid is found in blood plasma.
- Despite the large amount of water within the human body, large deviations can result in serious consequences.
- Acute reductions of 2% in body water (measured via changes in body mass) can result in altered cardiovascular system function, which can affect exercise performance at high temperatures.
- Acute reductions of 7% or more in body water increase the likelihood of collapse and risk of elevated blood sodium levels (**hypernatraemia**).
- Acute large increases in body water can lead to reduced blood sodium levels (**hyponatraemia**). This is known as water intoxication and can prove fatal.
- An individual is considered to be euhydrated when body water fluctuates by less than 0.22% at rest and in temperate environmental conditions. During exercise, or in high ambient temperatures, these thresholds increase to fluctuations of 0.48%.
- When body water increases above these thresholds, an individual is considered to be hyperhydrated.
- When body water decreases below these thresholds, an individual is considered to be hypohydrated.

Water balance

- Whole-body water balance is achieved when water intake matches water output.
- The main avenue for water intake is through the ingestion of fluids. Water is also ingested through foods containing water and a small amount of water is produced as a result of metabolic processes.
- The main avenue for water output while at rest and in temperate conditions is via elimination of urine. Water is also lost as a result of sweating, and additional small amounts are lost in faeces, expired air, and skin loss. While exercising or in hot environmental temperatures, water loss increases as a result of sweating.

Whole-body water and sodium balance

- Following a reduction in body water, the only way to increase body water levels is to ingest fluid. This is largely controlled by the sensation of thirst, although deviations in body water are not well detected by the thirst response. The thirst response is a complex interaction of physiological, psychological, and behavioural factors. The main physiological signals of thirst are a reduction in total blood volume and an increase in plasma osmolality.
- Urine output is largely regulated by hormonal factors. The two main hormones involved in this process are **vasopressin** (otherwise known as antidiuretic hormone) and aldosterone (the end product of the renin–angiotensin–aldosterone system). The main stimuli for release of these hormones are reductions in blood volume and increases in plasma osmolality. Consequently, the same physiological stimuli govern water loss and water intake.
- **Atrial natriuretic peptide** (or atrial natriuretic factor) is also involved in the regulation of urine output but to a lesser extent than arginine vasopressin and aldosterone.

Vasopressin

- Vasopressin (VP) is secreted from the anterior pituitary gland and is responsible for the reabsorption of water in the collecting ducts.
- Following secretion, VP binds to its receptor on the basal membrane of the collecting duct. This increases cytosolic concentrations of cyclic AMP and activates protein kinase A.
- The activation of protein kinase A leads to the insertion of **aquaporins** in both the apical and basal membranes. This, in turn, leads to an increase in membrane permeability to water resulting in an increase in water reabsorption.
- Secretion of VP results in greater water reabsorption. Consequently, a relatively low quantity of highly concentrated urine is formed.
- Urine osmolality is a commonly used measurement of hydration status and is effectively a measurement of the concentration of the urine. Urine osmolality can range from 50 to 1400mOsm/kg, whereas the osmolality of the plasma is very closely regulated between 275 and 300mOsm/kg. Therefore the kidneys are able to produce very small amounts of highly concentrated urine or very large amounts of low-concentration urine without causing much variation in plasma osmolality.
- Urine becomes concentrated while flowing through the medullary collecting ducts. The interstitial fluid around this area is very highly concentrated, which leads to large amounts of water reabsorption when VP is present. The ability to produce highly concentrated urine and conserve water is the result of the structure of the loop of Henle in the juxtamedullary nephrons.
- When entering the descending limb of the loop of Henle, the glomerular filtrate is at the same concentration as blood plasma. As discussed earlier, the descending limb is permeable to water, whereas the ascending limb is not. In addition, reabsorption of sodium chloride does not occur in the descending limb but does in the ascending limb (Figure 7.5).

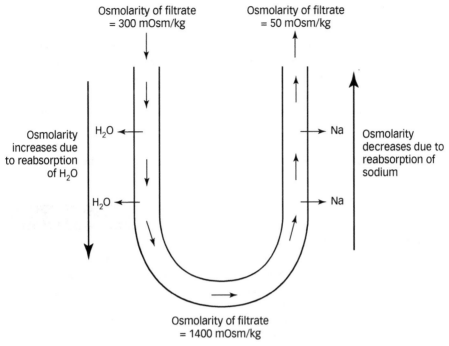

Osmolarity of filtrate = 300 mOsm/kg

Osmolarity of filtrate = 50 mOsm/kg

Osmolarity increases due to reabsorption of H_2O

H_2O

H_2O

Na

Na

Osmolarity decreases due to reabsorption of sodium

Osmolarity of filtrate = 1400 mOsm/kg

Figure 7.5 The counter-current multiplier system and the effect on filtrate osmolality

- As a result of the reabsorption of sodium chloride in the ascending limb, the osmolality of the interstitial fluid increases. Urea is also trapped within the interstitial fluid, leading to a further increase in osmolality. In turn, this leads to an increase in the amount of water that is reabsorbed in the descending limb due to osmosis. This effect is multiplied the further into the medulla the loop of Henle extends, leading to a maximum filtrate osmolality at the hairpin of the loop of Henle.

- While flowing through the ascending limb, the osmolality of the filtrate reduces due to the reabsorption of sodium chloride so that the osmolality of the filtrate is less than that of plasma when it arrives at the cortical collecting duct. In the presence of VP, water is reabsorbed until it has the same osmolality as plasma.

- When entering the medullary collecting duct, the presence of VP leads to more water reabsorption because of the high osmolality in the medullary interstitial fluid. This results in low volumes of very concentrated urine, which minimizes water loss and the potential effects of hypohydration.

- The main physiological stimulus for VP release is plasma osmolality. A change of 1mOsm/kg in plasma osmolality can lead to a change in the plasma concentration of VP of 0.41pmol/L. In addition, a change in plasma osmolality of 3mOsm/kg can lead to a change in urine osmolality of 250mOsm/kg, demonstrating the potent effect that plasma osmolality has on water output.

- A reduction in blood volume is also a stimulus for VP release. However, this is not as sensitive as changes in plasma osmolality. A reduction in blood volume of 8–10% seems to be an indicator of significant increases in plasma VP concentration.

- An applied example of this can be seen when an individual performs endurance exercise in a hot environment. During this exercise, water is lost as a result of evaporative heat loss, causing an increase in plasma osmolality and a reduction in blood volume. The increase in plasma osmolality leads to secretion of VP, which results in greater water reabsorption and the formation of low volumes of highly concentrated urine in an attempt to conserve body water.

- Diabetes insipidus is an inability to produce, or use, VP. Consequently, patients suffering from this condition produce large amounts of dilute urine even when hypohydrated.

Looking for extra marks?

Given the importance of VP in regulating urine production, it is beneficial in some research studies to measure circulating concentrations of this hormone. The most commonly used technique for measuring circulating concentrations of VP is enzyme-linked immunosorbent assay (ELISA), which can be costly. In such situations, free-water clearance can be calculated. Free-water clearance is the volume of water cleared from the plasma over a given time period. If the osmolality of the urine is equal to the osmolality of the plasma, free-water clearance is zero. A positive value indicates the clearance of water from the plasma and a negative value indicates a conservation of water. Although this is not a measurement of VP concentration, negative values suggest an increase in VP secretion and positive values indicate a decrease in secretion.

Free-water clearance can be calculated using a number of equations. For the simplest estimation, the following equation can be used:

$$\text{free-water clearance} = V \times \left[1 - \left(U_{Osm} / P_{Osm} \right) \right]$$

where V is urine flow rate (mL/min), U_{Osm} is urine osmolality (mOsm/kg), and P_{Osm} is plasma osmolality (mOsm/kg).

Suppose that an individual exercises for 1 hour in the heat and we obtain a urine sample and a blood sample at the end of this period. Urine flow rate would be equal to the volume of urine (mL) produced divided by time (min). If 100mL of urine is collected after 1 hour, the urine flow rate is 1.67mL/min. If we measure urine osmolality as 925mOsm/kg and plasma osmolality as 295mOsm/kg, we can calculate free-water clearance:

$$\text{free-water clearance} = 1.67 \times \left[1 - \left(925/295 \right) \right]$$
$$= 1.67 \times \left(1 - 3.14 \right)$$
$$= 1.67 \times \left(-2.14 \right)$$
$$= -3.57 \text{ mL/min}$$

This indicates that there has been an increase in VP secretion in order to conserve water during this time.

The renin–angiotensin–aldosterone system (RAAS)

- The primary function of the RAAS is to regulate blood pressure by altering the reabsorption of sodium, and therefore water, in the collecting ducts of the nephron.
- Following a reduction in blood pressure, detected by baroreceptors, or a reduction in plasma sodium concentration, detected by the macula densa, renin is secreted from the juxtaglomerular apparatus. Renin is an enzyme which catalyses the breakdown of angiotensinogen to angiotensin I.
- Angiotensin I is converted to the biologically active angiotensin II in the presence of angiotensin-converting enzyme.
- Angiotensin II acts on the adrenal gland and increases secretion of aldosterone. In addition, angiotensin II acts on arteriolar smooth muscle to cause vasoconstriction and increases secretion of VP.
- Aldosterone acts on receptors in the distal convoluted tubule and collecting ducts leading to an increase in the amount of sodium being reabsorbed within the nephron.
- More water is reabsorbed as a result of an increase in the osmotic gradient established by greater sodium reabsorption.
- A linear relationship between the degree of hypohydration, and consequent reduction in extracellular fluid volume, and aldosterone release has been observed.
- In addition to giving rise to a reduction in renin secretion, an increase in blood pressure also has a local effect on renal tubules, which results in a further reduction in sodium and water reabsorption.

Atrial natriuretic peptide

- The secretion of VP or synthesis of aldosterone cause an increase in water and sodium reabsorption, respectively, resulting in reduced urine output and water loss. Atrial natriuretic peptide (ANP) has the opposite effect by reducing water and sodium reabsorption, resulting in an increase in urine volume as well as sodium and water loss.
- ANP is synthesized within cells in the right atrium and is released in response to an increase in atrial distension. This is a direct result of an increase in extracellular fluid volume, which is often due to increases in plasma sodium concentration.
- ANP inhibits release of VP from the anterior pituitary gland and aldosterone from the adrenal gland. It also inhibits sodium, and therefore water, reabsorption within the collecting ducts.

7.4 ACID–BASE BALANCE

The regulation of blood pH is of the utmost importance. A relatively small increase or decrease in blood pH can lead to very serious side effects because of the effect that this may have on cellular function and enzyme activity. The renal system plays a central role in the acid–base balance.

- Blood pH is usually maintained between 7.35 and 7.45. An individual is considered to be in acid–base balance when the amount of hydrogen ions produced is the same as the amount secreted.
- An increase in pH above 7.45 is termed an alkalosis and a decrease in pH below 7.35 is termed an acidosis.
- Hydrogen ions are eliminated from the body via urine and expired air. Thus the renal system and the respiratory system have important roles in ensuring acid–base balance.
- The single most important factor affecting acid–base balance is the partial pressure of carbon dioxide in plasma. This is because carbon dioxide reacts with water to form carbonic acid. Carbonic acid then readily dissociates into hydrogen and bicarbonate ions. Consequently if carbon dioxide levels increase, pH decreases.
- An acidosis can be classed as either respiratory or metabolic. A respiratory acidosis is due to a reduction in breathing rate or efficiency, which leads to an increase in CO_2 and a reduction in pH. A metabolic acidosis is due to an increase in hydrogen ions brought about by an increase in metabolic processes. This occurs during high-intensity exercise for example.
- An alkalosis can also be defined as being respiratory or metabolic. A respiratory alkalosis occurs when the breathing rate increases, causing a reduction in CO_2 and an increase in pH. This may occur during a panic attack. Metabolic alkalosis is very rare, but is caused by an increase in bicarbonate ions. This condition can be caused by excessive vomiting.
- Blood pH is regulated by buffer systems, the respiratory system, and the renal system.

Blood buffering systems

- The purpose of a buffer is to temporarily regulate the pH of the plasma by the addition or removal of hydrogen ions.
- A **buffer** system tends to be a combination of a weak acid and a weak base which work together to release or add hydrogen ions, ensuring a stable blood pH.
- Amino acids provide important buffering systems in blood. Every amino acid contains an amino group and a carboxyl group at either end of the carbon chain. Carboxyl groups can release hydrogen ions in response to an increase in pH.

- In response to a reduction in pH, carboxyl groups can add a hydrogen ion and amino groups can also add a hydrogen ion to form amino ions. Overall, this causes an elevation in pH.
- Haemoglobin plays an important role as a buffer because carbon dioxide diffuses into red blood cells freely before being converted to carbonic acid and then hydrogen ions and bicarbonate. Bicarbonate is pumped out of the cell in exchange for chloride ions. However, the hydrogen ions bind to haemoglobin.
- The **carbonic acid–bicarbonate buffer system** plays a major role in regulating the pH of the extracellular fluid. As previously described, carbon dioxide and water react to form carbonic acid. This then dissociates into hydrogen ions and bicarbonate. This reaction is in equilibrium, meaning that a change in one constituent can lead to a change in any other. Consequently an increase in carbon dioxide leads to an increase in hydrogen ions and, conversely, a decrease in hydrogen ions leads to an increase in carbon dioxide. The equation for these reactions is:

$$CO_2 + H_2O \rightleftharpoons H_2CO_3 \rightleftharpoons HCO_3^- + H^+$$

- The carbonic acid–bicarbonate buffer system relies on normal functioning of the respiratory system and the availability of bicarbonate ions. Therefore a reduced ability to eliminate carbon dioxide via the lungs can lead to chronic problems with acid–base balance.

The role of the respiratory system

- As stated previously, the major determinant of blood pH is the partial pressure of carbon dioxide.
- A reduction in pH causes an increase in the partial pressure of carbon dioxide, which is detected by chemoreceptors and causes an increase in respiration. The increase in respiration causes the partial pressure of carbon dioxide to fall and pH returns to normal.
- An increase in pH causes a decrease in the partial pressure of carbon dioxide, which is, again, detected by chemoreceptors causing a reduction in respiration rate. This leads to an increase in the partial pressure of carbon dioxide and pH returns to normal.
- The mechanisms behind changes in the partial pressure of carbon dioxide causing changes in respiration rate are described in Chapter 6.
- Therefore the respiratory system plays a significant role in ensuring acid–base balance when an imbalance is caused by something other than altered respiratory function.

Looking for extra marks?

The Henderson–Hasselbalch equation can be used to determine the pH of a solution. In the case of the pH of blood, it can be used to show the relationship between pH and the constituents of the carbonic acid–bicarbonate buffering system. The equation is as follows:

$$pH = pK_a + \log\left[HCO_3^-\right] / \left[H_2CO_3\right]$$

where pK_a is the carbonic acid dissociation constant (6.1), $[HCO_3^-]$ is the plasma concentration of bicarbonate and $[H_2CO_3]$ is the plasma concentration of carbonic acid.

Under normal circumstances, the plasma concentrations of bicarbonate and carbonic acid are approximately 25mmol/L and 1.2mmol/L, respectively. Using the equation above, we can calculate the expected pH:

$$\begin{aligned} pH &= 6.1 + \log(25/1.2) \\ &= 6.1 + \log(20.8) \\ &= 6.1 + 1.31 \\ &= 7.41 \end{aligned}$$

This value falls within the normal range of blood pH of 7.35–7.45. However, if plasma carbon dioxide levels decreased, as would occur in the event of an increase in respiratory rate, the plasma carbonic acid concentration would reduce. If we assume that plasma carbonic acid concentration reduces to 1.0mmol/L, pH can be calculated as:

$$\begin{aligned} pH &= 6.1 + \log(25/1.0) \\ &= 6.1 + \log(25) \\ &= 6.1 + 1.4 \\ &= 7.5 \end{aligned}$$

This value is above the normal range of blood pH of 7.35–7.45, indicating an alkalosis.

The role of the renal system

- The purpose of buffers is to temporarily alter pH. However, in the case of an acidosis, hydrogen ions are not permanently removed from the body.
- Respiratory compensation provides a rapid response to deviations in acid–base balance resulting from non-respiratory causes.
- The renal system is a slower-acting mechanism which responds to changes in acid–base balance by altering bicarbonate secretion in the urine. This then leads to changes in plasma hydrogen ion concentration.
- When a deviation in acid–base balance occurs for reasons other than respiratory or renal factors, the regulation of pH is achieved by a combination of the respiratory and renal systems.

- When a deviation in acid–base balance occurs due to changes in respiratory function, the renal system is the only mechanism for restoration of pH. If the cause of the imbalance is altered renal system function, the only mechanism for restoration of balance is the respiratory system.
- The renal system increases or decreases the amount of bicarbonate secreted in urine. As can be determined from the carbonic acid–bicarbonate equation, the removal of bicarbonate from the plasma leads to an increase in hydrogen ion concentration. Conversely, the addition of bicarbonate to the plasma leads to a greater production of carbonic acid and a reduction in hydrogen ion concentration.
- Following an alkalosis, excess bicarbonate is excreted in the urine leading to an increase in plasma hydrogen ions such that pH returns to normal.
- Following an acidosis, bicarbonate secretion in urine is minimized. However, bicarbonate is also synthesized in the renal corpuscle and added to the plasma in order to return pH to normal.

Bicarbonate elimination and retention

- In situations of normal acid–base balance, plasma bicarbonate is filtered in the renal corpuscle and then effectively reabsorbed at various sections of the renal tubule.
- Within tubular cells, carbon dioxide reacts with water to form carbonic acid, which then dissociates into hydrogen ions and bicarbonate. The bicarbonate ion is diffused into the interstitial fluid, whereas the hydrogen ion is actively transported into the lumen where it reacts with the bicarbonate present in the filtrate resulting in the production of carbon dioxide and water. The carbon dioxide and water are then reabsorbed into the tubular cell and can begin the process again (Figure 7.6).
- As a result of this process, no bicarbonate is present in the filtrate and there is no bicarbonate loss via urine.

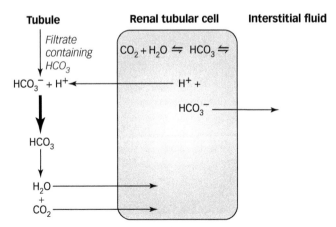

Figure 7.6 The reabsorption of bicarbonate under situations of normal plasma pH

- When hydrogen ion secretion in the tubular lumen exceeds the amount of bicarbonate within the filtrate, the excess hydrogen ions combine with other buffers such as hydrogen phosphate. Consequently, the bicarbonate ions that are formed from the dissociation of carbonic acid are moved into the interstitial fluid and ultimately the plasma, resulting in an increase in pH.
- In addition to this mechanism, bicarbonate is formed from the metabolism of glutamine by renal tubular cells. This leads to the production of ammonia, which is actively transported in the lumen and excreted in urine, and bicarbonate, which is diffused into the interstitial fluid and leads to an increase in plasma pH.
- In situations of alkalosis, the amount of bicarbonate in the filtrate exceeds the amount of hydrogen ions secreted from tubular cells and the metabolism of glutamine in renal tubular cells is relatively low. Consequently, a relatively large amount of bicarbonate is excreted in urine and relatively low amounts of bicarbonate are added to the plasma as a result of glutamine metabolism. The net effect of this is a reduction in plasma bicarbonate and a reduction in pH.
- In situations of acidosis, all the filtered bicarbonate is reabsorbed and new bicarbonate is added to the plasma as a result of hydrogen ions being buffered in the filtrate by hydrogen phosphate as well as an increase in glutamine metabolism within renal tubular cells. The net result of this is an increase in plasma bicarbonate and an increase in pH.

Check your understanding

Describe how the composition of urine differs from that of the glomerular filtrate. (*Hint: Consider the reabsorption and secretion processes within the nephron*)

Draw a diagram of a nephron and highlight the key functional structures. (*Hint: See Figure 7.2*)

Identify the four types of acid–base balance disorder and give an example of how each may occur. (*Hint: consider the interaction between renal and respiratory mechanisms*)

8 Gastrointestinal physiology

The digestive system is one of the most important systems in the human body. It is a hollow tract, approximately 7–9m long, which runs from the mouth to the anus with other associated organs, such as the liver and pancreas. While the main function of the digestive system is the digestion and absorption of nutrients, it also performs a wide variety of other functions including defence, transport, and elimination of waste products. In addition, it is the largest endocrine organ in the body.

Key concepts

- The digestive system has a number of functions other than the digestion and absorption of food.
- Ingested nutrients are broken down into their constituent parts by physical and chemical processes.
- The stomach acts as a reservoir allowing digestion to occur so that effective absorption can take place.
- All absorption of nutrients and water takes place in the intestine with the vast majority occurring in the small intestine.
- Carbohydrates are absorbed as monosaccharides using specific transport mechanisms.
- Fats are absorbed as part of mixed micelles.

continued

- Proteins are absorbed as either single or short-chain amino acids.
- The liver, gall bladder, and pancreas are associated organs that are centrally involved in the process of breaking down ingested nutrients.

8.1 THE DIGESTIVE SYSTEM

An understanding of the anatomy of the digestive system is key to understanding its function.

- The digestive system begins in the mouth where the process of nutrient **digestion** begins.
- From the mouth, the digestive system continues into the **oesophagus**, which acts as a transport mechanism for food and fluid to enter the stomach.
- The **stomach** has four regions: the **cardiac region**, the **fundus**, the **body**, and the **pyloric region**. The stomach acts as a reservoir for food and fluid, breaking down solid material, allowing further digestion to occur, and ensuring regulated movement of nutrients into the small intestine.
- The **small intestine** consists of the **duodenum**, the **jejunum** and the **ileum**. The main function of the small intestine is the absorption of nutrients.
- The **large intestine** consists of the **caecum** and the **colon**. The colon is separated into the ascending, transverse, descending, and sigmoidal sections of the colon. The main function of the large intestine is transport, with a limited amount of absorption (mainly water).
- The **rectum** and **anus** are responsible for the elimination of waste products.
- Associated organs of the digestive system include the **liver, gall bladder**, and **pancreas**. These are involved in the digestion of nutrients and fats in particular.

Histology

- All areas of the digestive tract have four layers: the **serosa**, the **muscularis**, the **submucosa**, and the **mucosa** (Figure 8.1).
- The depth of the layers differs between the different sections of the digestive system; however, the main difference observed is in the muscosal layer.
- The mucosal layer of the digestive system is the innermost layer and therefore is closest to the lumen (the inside space) of the tract.
- The mucosal layer consists of a layer of epithelial cells and loose connective tissue (called the **lamina propria**). A layer of smooth muscle cells, called the muscularis mucosa, is also present.
- The submucosal layer consists of dense connective tissue. This provides support to the mucosal layer and anchors the muscularis layer. The submucosal layer is also home to the **submucosal plexus** which forms part of the **intramural plexus**, and is strongly involved in regulation of function.

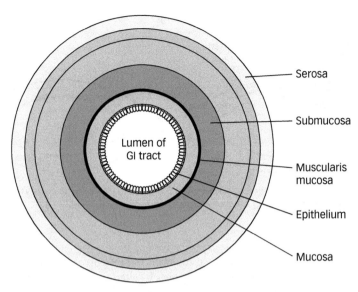

Figure 8.1 Cross-section of the digestive system showing the serosa, muscularis, submucosa, and mucosa

- The muscularis layer consists mainly of layers of circular and longitudinal smooth muscle fibres. It also contains the **myenteric plexus**, which forms the other part of the intramural plexus.
- The serosa consists of connective tissue and is the outermost layer of the digestive system wall.
- The digestive system has a rich nerve and blood supply.

Regulation of gastrointestinal function

- The regulation of gastrointestinal function is a complicated system, which involves the central nervous system, the **enteric nervous system**, and gut hormones, acting by circulating or paracrine mechanisms.
- Regulation of gastrointestinal function involves control of smooth muscle contraction and relaxation, which affect motility, blood flow, and other aspects of gastrointestinal function.
- Central nervous system activity comes from the autonomic nervous system. Sympathetic nervous activity usually uses noradrenaline as a neurotransmitter and results in relaxation of smooth muscle, a reduction in blood flow, and contraction of sphincters. Parasympathetic nervous activity usually uses acetylcholine as a neurotransmitter and results in contraction of smooth muscle, an increase in blood flow, and relaxation of sphincters (Figure 8.2).
- The enteric nervous system allows changes in gastrointestinal function to occur in response to changes in the local environment.
- The enteric nervous system consists of two enteric plexuses: the myenteric plexus and the submucosal plexus. Sensory neurons detect changes in the local

The digestive system

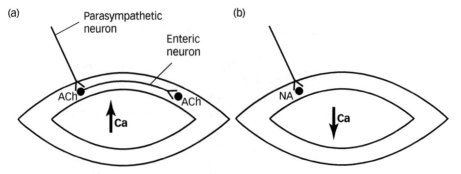

Figure 8.2 (a) parasympathetic and (b) sympathetic activation of digestive system smooth muscle and the effect on cytosolic calcium concentration (ACh, acetylcholine; NA, noradrenaline)

environment, such as temperature, osmolality, and motility. Motor neurons within the plexuses respond to these changes by altering the function of smooth muscle and other factors that affect gastrointestinal function. Interneurons are responsible for collating information from sensory neurons and communicating it to motor neurons. The main neurotransmitter used by enteric nerves is acetylcholine.

- The gastrointestinal system is the largest endocrine organ in the body and produces a large number of hormones that act as both autocrine and paracrine agents.

- Release of a number of hormones results in numerous effects on other body systems. Some examples are as follows.

  o **Ghrelin** is secreted mainly from the fundus of the stomach and acts on the hypothalamus, leading to an increase in appetite, as well as acting synergistically with growth hormone releasing hormone to increase secretion of growth hormone from the anterior pituitary.

  o **Peptide YY** is a hormone secreted from the ileum and colon, which results in a reduction in appetite. More information on the regulation of appetite is given in Chapter 10.

- Release of a number of hormones results in numerous effects on gastrointestinal system function. Some examples are as follows.

  o **Secretin** is a hormone that is released from the proximal small intestine and acts on the pancreatic ductal cells to increase bicarbonate secretion.

  o **Gastrin** is secreted from **G-cells** in the stomach and duodenum, and acts on enterochromaffin-like cells to stimulate histamine secretion, which acts on **parietal cells** to increase hydrochloric acid secretion.

  o **Motilin** is secreted from M-cells in the small intestine and acts on motilin receptors in the body of the stomach resulting in increased smooth muscle activity.

- The **migrating myoelectric complex** is a wave of smooth muscle contractions throughout the stomach and intestines that occurs when an individual is fasted. Its purpose is to enhance nutrient absorption as well as removing bacteria from the intestines. Motilin appears to have a significant role in inducing the migrating myoelectic complex. Food ingestion interrupts it.

The mouth and the oesophagus

- The process of digestion of nutrients begins in the mouth and includes both physical and chemical processes.
- The physical process of nutrient digestion may involve **mastication**. During this process, food is broken into smaller pieces, which assists in the digestive process by providing a greater surface area for enzymes to act on.
- The main enzyme involved in nutrient digestion in the mouth is salivary amylase.
- The oesophagus is guarded by the **upper oesophageal sphincter**. Sphincters usually allow one directional movement of material.
- The oesophagus acts as a transport mechanism for food and fluid to enter the stomach.
- The **lower oesophageal sphincter** is located in the lower part of the oesophagus at the entrance to the cardiac region of the stomach.
- The lower oesophageal sphincter is the centre of a variety of disorders including oesophageal achalasia and reflux oesophagitis.
- Oesophageal achalasia is a progressive disorder which results from an inability to relax the lower oesophageal sphincter. This may be due to problems with the enteric nervous system. As a result, ingested food and fluid remains in the oesophagus.
- Reflux oesophagitis results from a reduction in tone, or transient relaxations in, the lower oesophageal sphincter, or hiatus hernia. As a result, reflux of material from the stomach is possible, which may lead to erosion of the mucosa causing chest pain. Owing to the nature of this disorder, symptoms are usually worse at night or while in a supine position.

The stomach

- The stomach acts as a reservoir for food and fluid. This allows time for digestion to occur. Physical and chemical processes of digestion occur in the stomach. Physical mechanisms are the result of smooth muscle contraction and consequent movement of food. Chemical mechanisms are largely due to the secretion of hydrochloric acid and enzymes such as pepsinogen.
- No absorption of nutrients or water (except alcohol) occurs in the stomach.
- Parietal cells are responsible for the secretion of hydrochloric acid and intrinsic factor. These cells are mainly found in the body of the stomach (Figure 8.3).

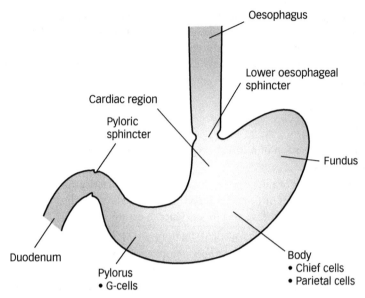

Figure 8.3 The stomach including the locations of key cells

- **Chief cells** are responsible for the secretion of pepsinogen (an enzyme involved in digestion of proteins). These cells are mainly found in the body of the stomach and are activated in response to increases in secretion of hydrochloric acid and gastrin.
- G-cells are responsible for the secretion of gastrin and are mainly found in the pyloric region of the stomach. They are activated in response to vagus nerve activity, and the secretion of gastrin results in increased secretion of hydrochloric acid from parietal cells and pepsinogen from chief cells.
- Mucus and bicarbonate are secreted from all cells in all regions of the stomach.
- **Rugae**, which are folds in an organ, can be seen in all areas of the stomach. This ensures that the stomach can expand in response to food and fluid ingestion.
- Following ingestion and swallowing of food, the smooth muscle of the body of the stomach relaxes (receptive relaxation). This allows an increase in stomach volume without a resultant increase in intragastric pressure.
- **Gastric emptying** is the term given to the process of movement of food (or **chyme**) or fluid from the stomach into the small intestine.
- In order for gastric emptying to occur, intragastric pressure has to increase while the **pyloric sphincter** (the sphincter that guards the entrance from the stomach to the duodenum) and duodenal smooth muscle have to relax.
- The increase in intragastric pressure is achieved through a contraction of stomach smooth muscle. This is the result of increases in central nervous system activity and the effects of numerous circulating hormones. Relaxation of the pyloric sphincter is achieved through changes in vagal nerve activity.

- The gastric emptying rate of solids is longer than that of liquids because of the digestion phase that is required in solids but not in liquids and also because of the reduced friction of liquids across the pyloric sphincter.
- Gastric emptying rate is affected by a number of factors including the volume of chyme present, nutrients, pH, osmolality, and exercise.

Hydrochloric acid secretion

- Hydrochloric acid is secreted from parietal cells, with the number of parietal cells determining the amount of hydrochloric acid secretion.
- Under resting conditions, parietal cells are pyramidal in shape with proton pumps that are not embedded in the cell membrane. Proton pumps are a key component of the mechanism of hydrochloric acid secretion; therefore they have to become embedded in the cell membrane before secretion can occur. The movement of proton pumps into the cell membrane is regulated by the actions of gastrin, acetylcholine, and histamine.
- Histamine acts on histamine receptors to increase the cytosolic level of cyclic AMP which, in turn, activates phosphokinase A.
- Acetylcholine and gastrin act on muscarinic and cholecystokinin B (CCK2) receptors resulting in an increase in cytosolic calcium concentration. This activates phosphokinase C.
- Hydrogen ions and bicarbonate are formed from the reaction of water and carbon dioxide. Bicarbonate is shuttled out of the cell in exchange for chloride. Hydrogen ions are pumped out of the cell in exchange for potassium and chloride diffuses into the lumen, resulting in hydrochloric acid.
- The epithelial layer of the stomach can withstand large amounts of hydrochloric acid, however, reduced barrier function or excess secretion may lead to ulceration.

The small intestine

- The duodenum is approximately 30cm long, the jejunum is approximately 2.5m long and the ileum is approximately 3.5m long.
- The main purpose of the small intestine is the digestion and absorption of nutrients, with the vast majority of this occurring in the duodenum and jejunum.
- The duodenum receives pancreatic juice and bile from the **ampulla of Vater** as well as chyme from the stomach.
- The end of the ileum is marked by the ileocaecal valve.
- The small intestine contains numerous folds. In addition to this, the mucosal layer of the wall projects into **villi**. Each villus contains a layer of epithelial cells that also project from the **brush border membrane**. This leads to a very large surface area for absorption to occur (Figure 8.4).
- At the lower end of several clustered villi is a **crypt of Lieberkuhn** where new epithelial cells are produced which replace cells shed from the villus tip. The dead cells are moved into the lumen which provides brush border enzymes that are involved in the digestive process.

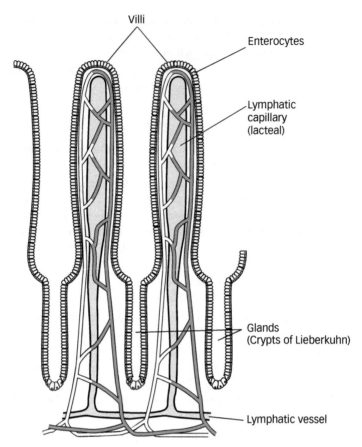

Figure 8.4 Small intestine histology showing villi, lacteal, and crypt of Lieberkuhn

- Each villus also contains a **lacteal** (a lymphatic capillary) and an extensive capillary network. Long chain fat is absorbed into the lacteal and into the lymphatic system.
- Carbohydrates and proteins are absorbed into the capillary network and are transported to the liver before blood is returned to the heart.
- The walls of the duodenum and ileum do not contain as many folds as those in the jejunum. This reflects the different roles of the sections of the small intestine, as most absorption, other than of water, occurs in the jejunum.
- The duodenum also contains duodenal glands, which are responsible for secreting mucus. Bicarbonate within the mucus counteracts the acidity of chyme from the stomach, ensuring protection of epithelial cells in the duodenum.
- The epithelial cells of the villus secrete a number of ions, including sodium and chloride, into the intestinal lumen. The increase in osmolality within the lumen leads to secretion of water into the small intestine. This water is reabsorbed further down the intestinal tract.

- Water movement within the intestine is the result of concentration gradients. If the contents of the intestinal lumen are less concentrated than the blood, water will move from the lumen into the blood. If the contents of the lumen are more concentrated than the blood, water will move into the lumen. The water that is moved into the lumen following ingestion of hypertonic solutions or when chyme is very concentrated is usually reabsorbed, but can lead to osmotic diarrhoea.

- Following movement into the duodenum, chyme is moved along the intestinal tract by peristaltic contractions. These are caused by pacemaker potentials that are exhibited in some smooth muscle cells within the small intestine. Activity of the enteric nervous system or the central nervous system, or the action of a number of hormones, can increase or decrease the rate of smooth muscle contraction. This results in increased or decreased intestinal motility.

Looking for extra marks?

Measuring gastrointestinal system function can be extremely challenging because of the invasive nature of many of the techniques commonly used. The method used to measure gastric emptying depends on whether solids or liquids are being investigated. Imaging techniques are often used for the measurement of gastric emptying. Scintigraphy includes ingesting a test meal which includes a radioisotope. A gamma camera is then used to picture where the tracer is in the stomach and the rate of gastric emptying can then be calculated. Other imaging techniques include magnetic resonance imaging and ultrasound. However, none of the imaging techniques are able to accurately quantify the volume of gastric secretions, which may be of interest in certain cases. Gastric emptying of liquids can be measured using these techniques and can also be measured using gastric aspiration. In this technique, a tube is placed at the base of the stomach before a test solution is ingested. The test solution contains a known concentration of phenol red. Gastric samples are taken at specific intervals before and after addition of known volumes and concentrations of phenol red. Total stomach volume, test meal volume, and the volume of gastric secretions can then be calculated.

Carbon-13 breath analysis is often used in clinical and research settings as an indirect measurement of gastric emptying. A substrate containing carbon-13 is added to the solution or food. This must empty from the stomach before being absorbed and is then rapidly metabolized by the liver before being exhaled in air. The ratio in breath samples of carbon-13 to carbon-12 is analysed and provides an indirect measure of emptying rate.

Intestinal absorption of solutions is also measured by intestinal perfusion. In this technique, two sampling ports are placed within the small intestine at a set distance apart. A solution containing a non-absorbable dye is infused and samples are taken from the second port. Net absorption can then be calculated. The ingestion of solutions containing deuterium oxide can also be used as a measure of overall fluid absorption. Deuterium oxide is added to a solution and blood samples are collected at regular intervals. These are analysed for deuterium concentration. The appearance of deuterium in the blood provides a measure of net uptake of the tracer, i.e. a combination of gastric emptying and intestinal absorption rate.

The large intestine

- The ileum and caecum are separated by the ileocaecal valve. This is not a true sphincter but exists to prevent backflow from the large intestine to the small intestine. It relaxes in response to contraction of ileal smooth muscle, which leads to movement of luminal contents into the large intestine.
- No folds or villi are present in the large intestine, which reflects its limited capacity for nutrient absorption.
- Sodium and chloride are reabsorbed in the large intestine with some resultant increase in water absorption. Potassium and bicarbonate are secreted into the large intestinal lumen.
- Very large numbers of bacteria colonize the large intestine. Some bacterial strains metabolize the contents of the intestinal lumen, which can lead to absorption of some short-chain fatty acids and vitamin K. Metabolism of some undigested compounds can lead to excess gas production and cause flatulence or diarrhoea.
- Smooth muscle contraction occurs at a slower rate in the large intestine than in the small intestine. This results in slower transit through this section of the digestive system.
- The internal anal sphincter is composed of smooth muscle. However, the external anal sphincter is composed of skeletal muscle and thus is under voluntary control.
- Following movement of faeces from the large intestine into the rectum, the increase in diameter causes contraction of smooth muscle in the rectum and relaxation of the internal anal sphincter. The increase in pressure would eventually result in the relaxation of the external anal sphincter and defecation. However, this is avoided due to the voluntary nature of skeletal muscle.

Looking for extra marks?

Inflammatory bowel diseases are chronic relapsing disorders that have no known cure. They tend to present in childhood and early adulthood. The incidence and prevalence of inflammatory bowel diseases has increased significantly in the last three decades, particularly in Western societies. The two major inflammatory bowel diseases are Crohn's disease and ulcerative colitis. Both types result in ulceration of the gastrointestinal system, which leads to abdominal pain and bloody diarrhoea. The disorders are characterized by periods of disease activity and periods of remission. Crohn's disease can affect any part of the digestive system, but is most commonly found in the ileum. The ulceration and inflammation that occurs in Crohn's disease can penetrate all layers of the wall. Ulcerative colitis is restricted to the colon or rectum. Ulceration and inflammation is mainly found in the mucosal layer of the wall. Potential complications of both types include intestinal perforation and an increased risk of intestinal cancer while Crohn's disease can lead to fistulas and obstruction. Complications are more likely in Crohn's disease, with the exception of the increased risk of intestinal cancer, which is more likely in ulcerative colitis. Because of the nature of the disorder, effects on other systems are possible. In particular, joint, eye and skin disorders are commonly found.

The gastrointestinal tract is home to an enormous number of micro-organisms. Tolerance to these organisms is essential for maintaining normal physiological function. Inflammatory bowel diseases are thought to be the result of a breakdown in this tolerance, which leads to uptake of bacteria or their products. This leads to changes in cytokine balance and immune function, which contribute to inflammation and ulceration. Changes in the composition of gut microflora may also contribute to the development of inflammatory bowel diseases. The development of inflammatory bowel diseases is a complicated area, but is likely to include a combination of environmental factors and genetics, which result in changes in immune function and lead to inflammation. There is some evidence to suggest that there are genetic factors that make an individual more susceptible to suffering from an inflammatory bowel disease. Most research in this area has focused on polymorphisms in genes, which encode a variety of proteins that are involved in ensuring microbial tolerance in the intestine. Owing to the rise in incidence of inflammatory bowel diseases in Western societies in recent years, much research has focused on the role of environmental factors in the development of these disorders. Inflammatory bowel diseases are more common in people of higher socio-economic status, people with sedentary occupations, and people working indoors. The mechanisms behind these observations are unknown. The most widely studied environmental factor in relation to inflammatory bowel disease is smoking status. Numerous investigations have suggested that smoking leads to an increased risk of developing Crohn's disease, but it seems to be protective against developing ulcerative colitis. In addition, patients diagnosed with Crohn's disease who smoke are more likely to relapse into activity of the disease and require surgery. The mechanism behind these observations is unclear. However, smoking has a variety of effects on cytokine response which may contribute to the development of these disorders. Changes in the composition of gut microflora are considered to be an important factor in the development of inflammatory bowel diseases. The most obvious way in which the composition of gut microflora can be altered is by changes in diet. Research in this area is subject to numerous methodological problems, but there is some evidence to suggest that diets high in sugar may be related to the development of inflammatory bowel disease. In addition, diets containing low amounts of omega-3 and/or high amounts of omega-6 fatty acids may lead to a greater likelihood of developing inflammatory bowel disease.

For further reading, please see Danese S, Sans M, and Fiocchi C (2004) Inflammatory bowel disease: the role of environmental factors. *Autoimmunity Reviews* 3: 394–400.

8.2 NUTRIENT ABSORPTION

One of the main functions of the digestive system is the absorption of ingested nutrients. The macronutrients present in the human diet are carbohydrates, fats, and protein.

Absorption of carbohydrates

- Carbohydrates are typically absorbed as **monosaccharides**. The monosaccharides that are found in the human diet are glucose, fructose, and galactose. They all have the chemical formula $C_6H_{12}O_6$ and are isomers of each other.
- Much of the ingested carbohydrate component of the human diet consists of disaccharides and polysaccharides. **Disaccharides** are two monosaccharides bound together following a dehydration synthesis reaction; for example, sucrose (table sugar) consists of a glucose molecule and a fructose molecule. **Polysaccharides** are multiple monosaccharides, such as starch or cellulose, formed as a result of multiple dehydration synthesis reactions.
- Following ingestion, disaccharides and polysaccharides have to be broken down via hydrolysis reactions to their constituent monosaccharides so that they can be absorbed in the small intestine.
- The digestion of polysaccharides begins in the mouth with salivary amylase and the process is continued by amylases secreted from the pancreas.
- Disaccharides are hydrolysed by enzymes secreted from the brush border membrane in the small intestine. Sucrase hydrolyses sucrose and lactase hydrolyses lactose. Lactose is the main sugar found in milk and consists of a glucose molecule and a galactose molecule.
- Lactose intolerance is a relatively common condition that results from a reduced ability to produce lactase. Consequently, when lactose is ingested, it is difficult to hydrolyse the disaccharide into glucose and galactose for absorption. Lactose is then moved into the colon where bacteria metabolize the sugar, causing an increase in gas. The increase in osmolality within the colon leads to water movement. As a result, the main symptoms of lactose intolerance are flatulence and diarrhoea. The main way to manage this condition is the removal of dairy products from the diet.
- Once disaccharides and polysaccharides are within the small intestine and have been hydrolysed to monosaccharides, there are four main barriers to absorption. There is a layer of mucus prior to the **apical membrane** of the enterocyte. Once carbohydrate crosses the apical membrane, the enterocyte can use the carbohydrate as an energy source before it has to cross the **basolateral membrane**.
- Crossing membranes requires the use of transporters. These transporters can be active (requires ATP) or passive (does not require ATP).
- Glucose and galactose are actively transported across the apical membrane by **sodium-linked glucose transporter 1 (SGLT-1)**. As the name suggests, SGLT-1 transports glucose while co-transporting sodium (Figure 8.5).

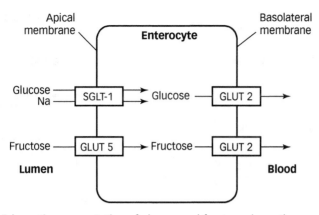

Figure 8.5 Schematic representation of glucose and fructose absorption

- Glucose is actively transported across the basolateral membrane by **glucose transporter 2 (GLUT2)**. GLUT2 appears to have a limited or maximum capacity for absorption and its activity is altered by insulin secretion and amino acids.
- Fructose is passively absorbed by a transporter, which is thought to be **glucose transporter 5 (GLUT5)**, on the apical membrane and by GLUT2 on the basolateral membrane.
- As passive transporters rely on concentration gradients, there appears to be a maximum capacity for absorption of fructose with acute ingestion of more than 40g resulting in malabsorption, although some individuals cannot tolerate significantly lower amounts than this. Co-ingestion of glucose seems to result in reduced malabsorption of fructose. Symptoms of fructose malabsorption resemble lactose intolerance so include flatulence and diarrhoea.

Absorption of fats

- The majority of dietary fat is in the form of triglycerides. This consists of a glycerol molecule and three fatty acids. The glycerol and fatty acid molecules have to be absorbed separately (see Figure 8.6).
- Other dietary sources of fats are phospholipids and cholesterol; however both these substances are present in bile that is secreted as part of the digestive process.
- The process of fat absorption begins in the mouth with lipases secreted from the tongue beginning the process of breakdown of triglycerides. This is continued in the stomach as a result of emulsification and the action of lipases from the tongue and stomach. The process continues in the small intestine with the secretion of bile from the liver and the packaging of fat into acid-coated droplets. This leads to a relatively large surface area and more efficient breakdown.
- Pancreatic lipase hydrolyses the fatty acids from triglycerides in preparation for absorption. This takes place in the upper part of the jejunum.
- Phospholipase, secreted from the pancreas, hydrolyses the fatty acids from phospholipids in preparation of absorption.

Nutrient absorption

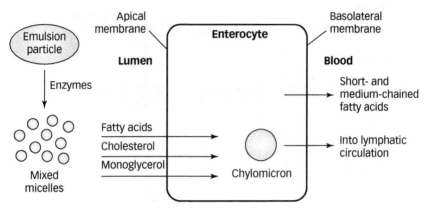

Figure 8.6 Schematic representation of fat absorption

- Cholesterol from the diet and that which is secreted in bile has to be non-esterified to be absorbed. About 50% of the cholesterol present in the intestine is absorbed and the rest is excreted. Esterified cholesterol is broken down into cholesterol by cholesterol esterase.
- Free fatty acids, monoglycerol, and cholesterol are packaged into **mixed micelles**, which are absorbed into the enterocyte via diffusion.
- Short- and medium-chain fatty acids are then directly absorbed into the portal circulation.
- Long chain triglycerides, cholesterol esters, and phospholipids are re-formed in the endoplasmic reticulum and the Golgi apparatus of the enterocyte.
- **Apolipoproteins** are added to the triglycerides, cholesterol esters, and phospholipids to form **chylomicrons**, which are absorbed into lymphatic vessels.
- Following absorption, apolipoproteins C and E are released from circulating **high density lipoprotein**, which results in activation of lipoprotein lipase and the breakdown of triglycerides from the chylomicron and storage of fat.
- The remaining chylomicron remnant is transported to the liver, where it is repackaged into **very low density lipoprotein.**
- When released into the blood, the very low density lipoprotein is converted to intermediate density lipoprotein as a result of activation of lipoprotein lipase and resultant storage of triglyceride.
- The intermediate density lipoprotein is converted to **low density lipoprotein** through further storage of triglyceride.
- **High density lipoprotein** transports cholesterol to the liver where it is eliminated.
- As a result of their functions, high circulating levels of high density lipoprotein and low circulating levels of low density lipoprotein are considered to be beneficial.

Absorption of proteins

- Proteins are chains of amino acids that can be of any length. Unlike carbohydrates, proteins do not necessarily have to be hydrolysed to single amino acids in order to be absorbed across the apical membrane of the enterocyte as short-chain proteins (two or gthree amino acids long) can be actively absorbed. Digestion and absorption of proteins occurs largely in the duodenum and jejunum.
- Protein found within the intestinal lumen can be from dietary protein ingestion or from secretions that occur in the mouth, stomach, and/or intestine.
- Pepsinogen is secreted from chief cells in the stomach and converted to its active form of pepsin. This begins the process of digesting proteins containing long chains of amino acids into shorter chains.
- This process is continued in the small intestine via the action of trypsin and chymotrypsin, which are enzymes found within the pancreatic juice. Enzymes from the brush border membrane also assist in this process. The end result of the digestive process is single or very short chains of amino acids.
- There are numerous transporters on the apical membrane which actively transport single amino acids. Some transporters have a higher affinity to certain amino acids than others. Co-transport of sodium occurs with single amino acid absorption across the apical membrane.
- Short chains of amino acids are actively transported with hydrogen ions. Once inside the cystosol of the cell, they are hydrolysed to their constituent single amino acids.
- There are numerous transporters on the basolateral membrane of the enterocyte which transport the single amino acids into the interstitial fluid. Again, this is an active process which requires ATP.

The liver, gall bladder, and pancreas

- The liver and pancreas have very important roles in the digestion of nutrients because of their ability to secrete bile, digestive enzymes, and bicarbonate. The gall bladder stores bile secreted from the liver.
- The pancreas lies behind the stomach and close to the duodenum. It has a distinctive shape which can be separated into the head (which is nearest to the duodenum) and the tail (which extends towards the spleen).
- Although the pancreas is often thought of as primarily an endocrine organ because of the secretion of **glucagon** and insulin from alpha and beta cells, the vast majority of the cells in the pancreas are exocrine in function. The pancreas is characterized by a series of ducts, which ultimately drain into the duodenum via the **duct of Wirsung** and the ampulla of Vater.
- The exocrine portion of the pancreas secretes pancreatic juice. This is a combination of ductal bicarbonate and digestive enzymes, such as pancreatic lipase (which assists in the digestion of lipids) and nucleases (which assist in the

breakdown of DNA and RNA). These enzymes would not act properly without the secretion of bicarbonate as the acidity of chyme is too high.

- The liver is located in the right upper quadrant of the abdomen and therefore is situated above the duodenum and near to the pancreas.

- The liver has numerous functions, in addition to those related to the digestive system, which are covered in other chapters.

- In terms of digestive system function, the liver is concerned with the production of **bile**. Bile contains bicarbonate (for neutralization of acidity of chyme), cholesterol, phospholipids, and bile salts. It plays an important role in the emulsification of lipids for absorption.

- The liver secretes bile into a duct system. The right and left hepatic ducts form the common hepatic duct. Bile can enter the **cystic duct**, which enters the gall bladder.

- The gall bladder is a small organ that sits underneath the liver. Its purpose is to store and concentrate bile prior to secretion into the duodenum.

- Smooth muscle contraction around the gall bladder, stimulated by CCK, is responsible for the movement of bile into the **common bile duct.**

- Bile can also be drained directly into the duodenum from the liver as the **common hepatic duct** also drains into the common bile duct.

- The common bile duct and the duct of Wirsung (from the pancreas) meet at the ampulla of Vater before draining into the duodenum.

- Movement of bile and pancreatic juice from the ampulla of Vater into the duodenum is controlled by the **sphincter of Oddi**. Relaxation of this sphincter, and consequent movement of bile and pancreatic juice into the duodenum also occurs in response to secretion of **cholecystokinin (CCK)**. CCK is secreted largely in response to ingestion of fats and proteins.

Looking for extra marks?

Irritable bowel syndrome is one of the most common functional gastrointestinal system disorders. It affects approximately 10% of Western populations and results in significant health costs. Historically, irritable bowel syndrome has been thought of as a non-organic disorder, i.e. a disorder which cannot be easily observed or measured as there is no clear anatomical or biochemical cause. However, recent advances have suggested that it may be an organic disorder. Irritable bowel syndrome presents as abdominal pain, which is relieved upon defecation, and a change in bowel habits. It is often described as being constipation predominant or diarrhoea predominant, however, this is a simplistic characterization as individuals often have periods of both constipation and diarrhoea. The most troublesome symptom in many people is bloating. Historically, irritable bowel syndrome has been diagnosed by elimination of other disorders, but diagnosis has been advanced in recent times. Treatment of irritable bowel syndrome is often ineffective and most commonly includes management of symptoms. Antispasmodics and anticholinergics are prescribed in some cases to reduce muscle spasms. Dietary approaches and psychological treatments are also important. Irritable bowel syndrome exhibits a placebo response rate of approximately 40%.

Possible reasons for the development of irritable bowel syndrome have been suggested. As pain is one of the main symptoms, a possible explanation is hypersensitivity of the visceral organs (termed visceral hypersensitivity). There is some evidence to suggest that sufferers from irritable bowel syndrome exhibit a degree of visceral hypersensitivity and also that they activate brain centres involved in distraction more than those without irritable bowel syndrome. Previous infection with Campylobacter or Salmonella bacteria is associated with the likelihood of developing irritable bowel syndrome. There is evidence, though inconclusive, to suggest that there is a degree of intestinal serotonin dysregulation in those who suffer from irritable bowel syndrome. Serotonin is involved in gastric and intestinal motility. Consequently, increased or decreased levels of serotonin may result in changes in intestinal smooth muscle contraction and lead to constipation or diarrhoea. There is some evidence that there is a genetic component to the development of irritable bowel syndrome, but the genes that may be involved have not been well researched. Probiotics have been the focus of a great deal of research, and some strains have been shown to be effective in reducing some symptoms of irritable bowel syndrome although others have been shown to be ineffective. This remains highly controversial.

For further reading, please see Talley NJ (2006) Irritable bowel syndrome. *Internal Medicine Journal* 36: 724–8.

Check your understanding

Describe why proton pump inhibitors are effective in treating a number of upper gastrointestinal disorders. (*Hint: Consider how hydrochloric acid is secreted from parietal cells and the role that histamine has in this process*)

Describe why a combined glucose–fructose solution may result in a greater rate of water and carbohydrate absorption than a glucose-only or fructose-only solution. (*Hint: Consider the mechanisms behind the absorption of these carbohydrates*)

Describe why the use of pharmacological agents that inhibit pancreatic lipase may be beneficial for weight loss. (*Hint: Consider the roles of these enzymes in the process of fat absorption*)

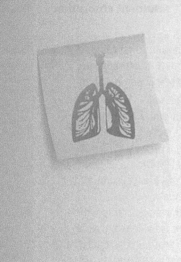

9 Reproduction

Reproduction is the process of producing offspring from parents. Genetic material from each of the parents is transmitted to the offspring, who therefore shares this material but is not genetically identical to either parent. Genetic 'fusion' occurs when a male sperm cell fertilizes a female ovum or egg cell. The fertilized egg then undergoes a series of phenomenal changes over the next 9 months, resulting in the birth of an offspring. This chapter considers these processes.

Key concepts

- Reproduction is a closely regulated process.
- Male gamete formation is a continuous process.
- Female gamete production is a cyclical process and only occurs for about 30–35 years in females.
- The same hormones of the hypothalamo-pituitary axis (GnRH, FSH, LH) control gamete formation in both males and females.
- Fertilization completes the meiotic division of ova and results in the formation of a zygote, which will ultimately divide and grow to produce a fetus. Nutrition is provided to the fetus and waste products from it are removed by the placenta.
- Maturation of the fetal adrenal glands is thought to trigger parturition.

9.1 MALE REPRODUCTIVE PHYSIOLOGY

The male reproductive tract is responsible for production of sperm cells and the deposition of semen into the female during sexual intercourse.

Organization of the male reproductive tract

- The male reproductive tract is external and consists of a **penis** and a pair of **testes** enclosed within the scrotum (Figure 9.1).
- The testes are **gonads** i.e. the structures responsible for the production of **sperm**—a process called **spermatogenesis**. They are also the primary source of the male sex hormones, known collectively as androgens. The best known example of an androgen is **testosterone**. A cross-section through a testicle is shown in Figure 9.2.
- The testes are composed of a series of coiled tubes called **seminiferous tubules**.
- There are two important cell types in the seminiferous tubles: **Sertoli cells**, which are responsible for sperm production, and less numerous **Leydig cells**, which are responsible for testosterone production and release. Leydig cells are sometimes called interstitial cells.
- Individual seminiferous tubules drain into structures called the rete testes.
- The rete testes drain into a highly coiled tube called the **epididymis**, which in turn drains into the **vas deferens**. The vas deferens from each testicle dilates and forms a structure called the ampulla of the vas deferens. The ampulla merges with the outflow from the seminal vesicle to form a structure called the **ejaculatory duct**.

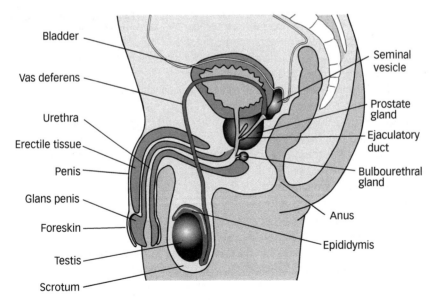

Figure 9.1 General organization of the male reproductive tract

Male reproductive physiology

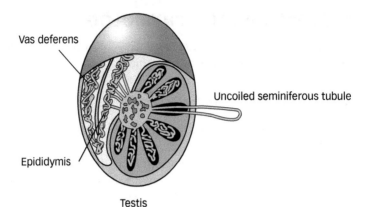

Figure 9.2 Cross-section through a testicle

- The ejaculatory ducts enter the prostate gland, merge together, and enter the urethra, which allows sperm and the secretions of various glands to be released from the penis—the process of ejaculation.
- Whilst sperm are produced in the testes and mature as they move through to the epididymis, it is **semen** which is ejaculated.
- Semen is a mixture of sperm and secretions of the seminal vesicle and prostate gland. These secretions provide nutrients for the sperm cells, are alkaline (to buffer the acidic secretions of the female reproductive tract), and contain anticoagulants. A typical ejaculatory volume is 2–5mL, 80% of which is made up from glandular secretions.

Looking for extra marks?

The role of the testes is to produce sperm; the role of the penis is to introduce them into the female reproductive tract. In order to accomplish this, the penis must become erect. The penis contains erectile tissue—essential venous sinusoids which have the ability to become engorged with blood. The erectile tissue is composed of two corpus cavernosae and a single corpus spongiosum. Parasympathetic activity to the pudendal artery results in increased blood flow to the penis. Venous outflow is unaltered and the erectile tissue becomes engorged with blood. As this continues it occludes the emissory veins, further reducing venous outflow; hence the penis becomes erect. During intercourse, the vas deferens displays contractile activity—this helps to move sperm from here into the urethra, resulting in ejaculation.

In addition to androgens, the testes also secrete other essential chemical substances. These include Mullerian-inhibiting hormone (MIH). MIH is essential for the regression of the Mullerian ducts in the fetus. If the Mullerian ducts were allowed to develop, they would form a uterus and Fallopian tubes.

Spermatogenesis

- The production of functional sperm is known as spermatogenesis. This begins at puberty and continues, although at a reduced efficiency, until death. Once spermatogenesis is initiated, a male may produce 200–250 million sperm a day.
- The first step in spermatogenesis is the production from diploid germ cells of cells called **spermatogonia**. Diploid means that the cells have a full complement of chromosomes, i.e. 23 pairs comprising 22 pairs of autosomes and one each of the X and Y chromosomes (the so-called sex chromosomes).
- Spermatagonia undertake two further mitotic cell divisions to form primary spermatocytes, which are still diploid.
- They then begin to undergo meiotic cell divisions to form structures called secondary spermatocytes.
- Secondary spermatocytes undergo a further meiotic cell division to form spermatids. These are now haploid cells. They contain one of each pair of autosomes and one of the sex chromosomes – making 23 chromosomes in total.
- The final stage in the production of sperm is the maturation and formation of sperm cells from the spermatids – a process called spermiogenesis.
- As the development from spermatagonia to sperm occurs, so the cells move through the Sertoli cells towards the lumen of the seminiferous tubules (Figure 9.3). It takes

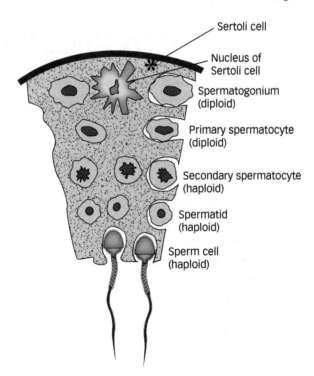

Figure 9.3 Cross-section through a seminiferous tubule

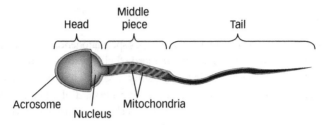

Figure 9.4 Structure of a typical sperm cell

about 70 days to produce sperm cells from spermatagonia. The structure of a typical sperm cell is shown in Figure 9.4.

- Sperm consist of three regions: head, mid-piece, and tail.
- The head contains the nucleus (which contains half the full chromosome number). Within the head is the **acrosome**. This contains hydrolytic enzymes whose function is to allow the sperm to penetrate the egg at **fertilization**.
- The mid-piece is characterized by spirally arranged mitochondria.
- The tail of the sperm is motile, powered by the ATP produced in the mitochondria of the mid-piece. The tail is a single flagellum and is identical to the 9 + 2 flagella seen in other organisms.
- Sperm cells lack other intracellular organelles.

Looking for extra marks?

The testes have a blood–testes barrier, analogous to the blood–brain barrier, which is formed by a group of cells in the seminiferous tubules called nurse cells. This ensures that developing sperm cells are not exposed to potentially harmful substances in the plasma. These include cells of the immune system, which would recognize developing sperm cells as 'non-self'.

Hormonal control of testicular function

- As indicated earlier, one of the roles of the testes is to produce the male sex hormones (androgens)—primarily testosterone.
- Testosterone has a number of important functions:
  - it stimulates sperm production
  - it promotes the development of male secondary sexual characteristics at puberty
  - its anabolic effects include increased protein synthesis and muscle growth— hence its use as a banned substance in sport
  - via its effect on the CNS, it increases libido.
- Testosterone is produced by the Leydig cells in the testes.
- The vast majority enters the plasma; however, some enters the seminiferous tubules where it binds to androgen-binding protein and stimulates sperm production.

- It is important that levels of testosterone are regulated, ensuring that male reproductive function is optimized.
- Hormonal control of testicular function is achieved by the hypothalamo-pituitary axis.
- Gonadotrophin releasing hormone (GnRH) is synthesized and released from neurons in the hypothalamus.
- It enters the portal blood vessels and travels to the anterior pituitary gland where it promotes the release of follicle stimulating hormone (FSH) and luteinizing hormone (LH).
- LH targets Leydig cells, where it promotes the release of testosterone.
- FSH targets Sertoli cells, where it promotes the release of androgen-binding protein. It also promotes the release of inhibin and the enzyme aromatase, which promotes the conversion of testosterone to the female sex hormone oestradiol.
- The released testosterone inhibits the release of further LH by reducing activity in both the hypothalamus and the anterior pituitary gland.
- Increased levels of inhibin and aromatase inhibit activity in the hypothalamus and anterior pituitary gland and reduce FSH secretion.

Looking for extra marks?

Testosterone is produced from acetic acid and cholesterol via a series of intermediaries, including progesterone (a female sex hormone). In order to exert its peripheral effects (e.g. influencing muscle mass), it must enter the plasma and circulate. However, as it is a lipid-soluble compound, it does this by binding to transport proteins in the plasma (e.g. gonadal steroid binding globulin). It leaves this transporter to enter target cells (e.g. muscle), where it binds to intracellular receptors. Some of the testosterone that enters the cell is converted to a compound called dihydrotestosterone (DHT). Some target cells are more responsive to DHT than to testosterone; however, they both have the ability to bind to intracellular receptors. It is known that testosterone levels decrease with age, and that this reduction may be correlated with a number of features of ageing (e.g. decreased muscle mass).

9.2 FEMALE REPRODUCTIVE PHYSIOLOGY

The female reproductive tract is responsible for the production of ova and the development of the fetus. Following birth there is a variable period when the mother is the sole source of nutrients for the baby.

Organization of the female reproductive tract

- In contrast with the male reproductive system, the female reproductive system is internal (Figure 9.5).

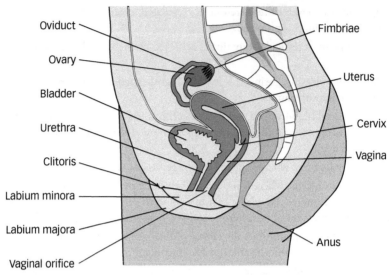

Figure 9.5 General organization of the female reproductive tract

- Paired **ovaries** lie within the pelvis—these are the gonads of the female reproductive tract and responsible for ensuring that, on a cyclical basis, an ovum is prepared for fertilization.
- Functioning with the ovaries are a pair of **Fallopian tubes**, which originate from the **uterus**.
- Sperm deposited in the vagina travel to the distal regions of the Fallopian tubes where fertilization of the ovum occurs. They then allow the fertilized egg to travel back to the uterus where it will initially implant in the uterus prior to the formation of the placenta.
- The uterus contains two important regions. The innermost lining called the **endometrium** undergoes cyclical changes to ensure that a fertilized egg has an initial location for development to proceed prior to the formation of the placenta. The uterus also contains smooth muscle—the **myometrium**. Activity in this muscle expels the fetus at the end of pregnancy.
- The cervix is the start of the birth canal. It is linked to the **vagina**, allowing deposition of sperm during sexual intercourse and also forms the birth canal during childbirth.
- The opening to the vagina is protected by the vulva, which consists of folds of tissue called the labia majora and the labia minora.

Ovary function and oogenesis

- Unlike the male reproductive system, in which sperm are produced continually, egg or ovum production in females occurs in a cyclical (i.e. monthly) fashion. Also unlike males, where sperm production continues from puberty until death, ovum production occurs from puberty until the onset of the menopause.

- The ovary at birth contains all the ova a woman will ever possess. These ova are produced from stem cells called **oogonia**, which are diploid.
- During the latter stages of fetal development, the oogonia begin the process of meiosis. By the time of the birth, this results in the formation of structures called primordial follicles. Primordial follicles consist of an ova surrounded by a layer of cells called the follicle cells. However, this process of meiosis does not proceed to completion; the ova within the primordial follicles remain suspended in the prophase stage of the first meiotic division until puberty when, on a cyclical basis, an individual ovum completes its meiotic division and the formation of a haploid cell.
- The first meiotic division is completed after puberty under the influence of FSH, but the final meiotic division is only completed after an ovum has been fertilized.
- At puberty a number of primordial follicles are allowed to begin to complete the process of further meiotic division. However, only one follicle will complete the process each month and the rest will degenerate—a process called **atresia**.
- One follicle is allowed to complete its development once a month, although in most women this duration varies between 25 and 35 days. This is the ovarian cycle, which is more commonly referred to as the **menstrual cycle**. Theoretically, the term menstrual cycle refers to cyclical changes in the endometrium.
- A typical ovarian cycle is considered to begin on the first day of bleeding of the menstrual cycle.
- During the first couple of days, the primordial follicle increases in size and forms the preantral follicle. During this phase, cells of the follicle begin to express receptors for both LH and FSH.
- Under the influence of FSH and LH the preantral follicle develops into an antral follicle. Antral follicles are also known as Graffian follicles. As this process proceeds, cells within the follicle begin to secrete **oestrogens**, particularly **17-β-oestradiol**.
- Towards the end of this phase of follicular development, there is a further rise in the number of LH receptors (and also LH levels). As a consequence, the follicle ruptures and the oocyte is released. This is known as ovulation. During this phase the ovum completes its first meiotic division. This occurs after about 14 days, i.e. mid-cycle.
- As ovulation occurs, activity of cilia in the distal ends of the Fallopian tubes—the fimbriae—pulls the ovum into them.
- The empty follicle left behind after ovulation is called the **corpus luteum**. In the event of pregnancy occurring, the corpus luteum assumes an endocrine role and maintains the endometrium in preparation for implantation by the fertilized egg.
- The corpus luteum begins to secrete large amounts of **progesterone**.
- If the ovum released at ovulation remains unfertilized, the corpus luteum degenerates. This occurs 12–14 days after ovulation and marks the end of a single ovarian cycle.
- The first half of the ovarian cycle is known as the follicular phase and is dominated by oestrogen; the second half is known as the luteal phase and is dominated by progesterone.

Cyclical changes in the endometrium – menstrual cycle

- It is essential that as the development of ova in the ovaries proceeds each month, there are changes in the endometrium which, in the event of fertilization occurring, will become the first 'home' for the fertilized egg.
- The cyclical changes which occur in the endometrium are known as the menstrual cycle.
- The menstrual cycle begins on the first day of bleeding (menses).
- During the follicular phase of the ovarian cycle, the high oestrogen levels prepare the reproductive tract for possible fertilization and subsequent implantation of the zygote.
- Under the influence of oestrogens, activity in the cilia of the fimbrae increase. The fimbrae are the most distal regions of the Fallopian tubes. This ensures that the ovum released from the follicle is 'captured' and is transferred back along the Fallopian tubes towards the uterus.
- At the same time, the endometrium begins to proliferate and, as it does, there is an increase in its vascularization (spiral artery development) and an increase in its secretions. During this time the endometrium also becomes primed to respond to progesterone in the latter half of the cycle.
- This phase of the menstrual cycle is known as the **proliferative phase**.
- In the second half of the menstrual cycle (which coincides with the luteal phase of the ovarian cycle) the endometrium is fine-tuned in preparation for pregnancy.
- The development of the spiral arteries within the endometrium continues.
- Likewise, the secretions in the endometrium continue. However, they change from the watery secretion seen in the proliferative phase to one which is now rich in nutrients.
- This phase of the menstrual cycle is known as the **secretary phase**.
- If fertilization does not occur, the corpus luteum degenerates. Consequently, the source of progesterone, which is maintaining and developing the endometrium, is lost. Thus the endometrium is lost—menstruation occurs and a new ovarian and menstrual cycle begins.

Looking for extra marks?

Hormonal contraceptives work by influencing the levels of hormones during the ovarian and menstrual cycle. Some of these contraceptives contain both oestrogen and progesterone; they work by preventing ovulation. A second type contains progesterone only; this acts to modify the local environment of the reproductive tract to make it less receptive to sperm and fertilization.

Control of female reproductive physiology

- It is essential that both the ovarian and the menstrual cycle run parallel to each other. Therefore it is imperative that the ovarian and endometrial functions are closely controlled.
- The cycle begins on the first day of menses. At this time, under the release of GnRH, FSH is released, which promotes follicle development. Shortly after this, LH secretion begins to rise.
- During the first half of the cycle, cells in the follicle begin to express receptors for LH. Plasma levels of oestrogen begin to increase as the follicle starts to synthesize and release oestrogen.
- As the level of oestrogen continues to rise, rather than producing a negative feedback influence on the anterior pituitary hormone, it produces a positive feedback effect—particularly on LH production.
- This produces a so-called LH surge, which results in the occurrence of ovulation.
- After this, the follicle begins to produce progesterone, and so the levels of progesterone rise while the levels of oestrogen begin to reduce.
- The positive feedback that oestrogen had on FSH and LH secretion is removed and negative feedback is reasserted. The result of this is that the FSH and LH levels drop.
- Oestrogen levels may continue to rise, but there is no second surge in LH—the secretion of progesterone appears to inhibit this.
- If fertilization fails to occur, menstruation starts and the FSH and LH levels begin to rise again as a new cycle starts.

9.3 FERTILIZATION

Fertilization is the union of sperm and ova to produce a zygote, which will ultimately develop into a fetus.

- **Fertilization** represents the union of a single sperm with a single ovum to produce a diploid **zygote**.
- Sexual intercourse results in the deposition of 2–5mL of semen, which contains 200–300 million sperm cells, in the female reproductive tract. Sperm are deposited in the vagina and must travel to the distal ends of the Fallopian tubes in order to fertilize the single released ovum.
- Fertilization has a narrow time window in which to occur—sperm only remain viable in the female tract for about 36-48 hours after ejaculation and ova only remain viable for 12–24 hours post-ovulation.
- Once deposited in the female, sperm must undergo a process called **capacitation**. Capacitation is a calcium-dependent process—the membrane at the head fuses with the membrane that forms the acrosome. Capacitation only occurs once the sperm have been deposited in the vagina.

- When a sperm meets an ovum the acrosome reaction occurs. This results in the release of hyaluronidase from the acrosome. This is a hydrolytic enzyme, which digests the follicular cells surrounding the ovum. The sperm then fuses with the membrane of the ovum to form a zygote.
- This triggers the ovum to complete its second meiotic division, which takes about 2 hours. If this fails to occur, there is a risk that the fertilized ovum may display triploidy, i.e. three sets of chromosomes.
- After this polyspermy must be blocked, i.e. no other sperm must be allowed to fuse with the ovum. The mechanism by which this occurs is unclear.
- The final step in the process of fertilization is that the fertilized ovum must now prevent the regression of the corpus luteum and thus the loss of the endometrium. The fertilized egg does this by secreting a hormone called **human chorionic gonadotrophin (hCG)**. The zygote then implants in the endometrium and the gestation period—9 months of growth and development—begins.

Looking for extra marks?

hCG can be found in the urine of a pregnant woman about 14 days after fertilization. This means that it can be used as a test for pregnancy—its detection by a colorimetric method is the basis of home pregnancy testing kits.

9.4 FORMATION OF THE PLACENTA AND DEVELOPMENT OF THE ZYGOTE

The placenta is a complex structure which allows the passage of nutrients and waste materials between the fetus and its mother.

- The period of gestation, from formation of the zygote to birth, is about 9 months. In order for this to occur, there must be an adequate delivery of oxygen and nutrients and removal of waste products. This is achieved by the formation of a structure called the placenta which allows maternal blood to exchange products with the blood supply of the developing fetus. The formation of the placenta is a complex process.
- Immediately after fertilization, the zygote begins to undergo mitotic cell divisions—a process called cleavage, which results in the formation of a structure called a **blastocyst**. This is completed about 7 days after fertilization.
- The blastocyst is a hollow ball of cells (Figure 9.6). The outer layer is called the **trophoblast** and will go on to contribute to the formation of the placenta. There is also a collection of cells gathered at one end called the inner cell mass, which will eventually become the embryo.
- The region of the trophoblast nearest to the inner cell mass makes contact with the endometrium and, through the release of enzymes, it burrows into the endometrium. As this happens, the inner cell mass detaches and pulls away from the trophoblasts. This creates a space that will ultimately form the amniotic cavity.

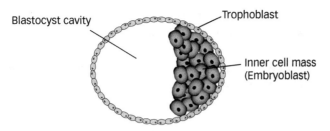

Figure 9.6 Structure of a blastocyst

- About 12–14 days after fertilization, a process known as **gastrulation** occurs. This results in a 'rearrangement' of the cells and the formation of three embryonic germ layers—endoderm, mesoderm, and ectoderm.
  - The endoderm will eventually form structures which form parts of the reproductive, urinary, and digestive systems amongst others.
  - The mesoderm will form the skin and skeletal system and all of the muscular and cardiovascular system.
  - The ectoderm forms all of the nervous system and parts of the endocrine, respiratory and digestive systems.
- The next 10 weeks sees the occurrence of a process of embryogenesis. This sees the formation of a structure which has a definite body shape and also of the internal organs. The period of time to the end of embryogenesis coincides with the first third, or trimester, of pregnancy.
- The second and third trimesters correspond to further growth and development of the embryo, which by this time is known as a fetus.
- By the end of a typical 9 month pregnancy, the fetus will weigh about 6–8 pounds and will have functioning physiological systems, ready for when it leaves the confines of its mother and enters the world.

Looking for extra marks?

Whilst the fetus is developing inside its mother, considerable burdens are placed on her normal physiology. For example, there is an increase in circulating blood volume, nutrient requirements, and renal function. The burden is less during the early stages of the pregnancy, but increases as the pregnancy proceeds. One of the most significant functional changes seen in pregnant women is the development of the mammary glands in preparation for lactation following the birth.

9.5 PARTURITION

Parturition or birth is a process which involves the expulsion of the fetus from the mother. Current thinking suggests that the fetus plays a significant role in initiating this process.

Lactation

- Parturition is the expulsion of the fetus from its mother to the external world.
- It is characterized by dilation of the cervix and in increase in the contractility of the uterus.
- What triggers parturition is not entirely clear, but it is thought that the fetus itself gives the signal for the process to begin.
- It is thought that the process is triggered by an increase in the production of fetal **cortisol**.
- In turn, cortisol promotes the conversion of progesterone to oestrogens in the placenta. Oestrogens increase the contractile activity of the uterus and also make it more sensitive to other substances, which increase contractile activity (e.g. oxytocin and some of the prostaglandins).
- The increased levels of oestrogen promote the synthesis and release of prostaglandin F2α by the placenta. This increases contraction of the uterus.
- The increased contractility of the uterus increases the release of oxytocin from the posterior pituitary gland, which in turn initiates further contraction. The release of oxytocin and the process of parturition is subject to positive feedback.
- Eventually, contraction of the uterus results in expulsion of the fetus from the mother.
- Following expulsion of the fetus, contractions of the uterus continue to occur. This results in rupture of the connections between the placenta and the endometrium, and the placenta is expelled as the afterbirth. Although this is associated with blood loss, the contractile activity of the uterus and increased circulating blood volume of the mother mitigate this.

9.6 LACTATION

Lactation describes the production of milk in breast tissue. This is the initial source of nutrients for the newborn baby.

- Until birth, the fetus has obtained all nutrients from its mother via the placenta.
- In the immediate period following birth, the baby receives nutrients from breastmilk produced by its mother.
- Development of the breast begins during pregnancy to ensure that the considerable nutrient needs of the baby are met.
- Most of the development of breast tissue is completed approximately half way through the pregnancy.
- Breast tissue consists of ducts known as lactiferous ducts, which terminate at their distal end in structures called alveoli. At their proximal end the lactiferous ducts open into the nipple region allowing milk to leave the breast.
- During pregnancy the ducts develop and grow under the influence of a variety of hormones, including progesterone and other placental hormones.
- The alveoli are the sites of milk production. Associated with the alveoli are contractile cells called myoepithilial cells, which are responsible for moving milk

into the lactiferous ducts. The lactiferous ducts allow milk to be ejected from the nipple

- One of the most important hormones responsible for milk production is the anterior pituitary hormone **prolactin**.
- Following parturition and loss of the placenta, there is a sudden loss of the steroids (oestrogens and progesterone) produced by the placenta. This loss of steroids allows prolactin to exert its effects on breast tissue – hitherto, breast tissue is unresponsive even to the high of levels of prolactin seen during pregnancy.
- Following birth, prolactin continues to be the hormone responsible for driving milk production.
- An infant suckling at the breast ensures that prolactin continues to be produced. Suckling produces a reflex release of prolactin. It does this in an indirect manner in that it appears to produce a drop in dopamine (prolactin inhibiting hormone) from the hypothalamus.
- The loss of a suckling stimulus (e.g. when a baby is weaned onto other food sources) results in the suppression of milk formation and breast tissue reverts to its non-pregnant, non-suckling state.

Check your understanding

What are the significant differences between male and female gamete production? (*Hint: consider the nature, duration, and time span of formation*)

What are the potential adverse effects of abuse with anabolic steroids in males? (*Hint: think about the normal roles of testosterone*)

Why are drugs which interfere with ovarian function effective contraceptives? (*Hint: think about the normal regulation of activity in the ovary*)

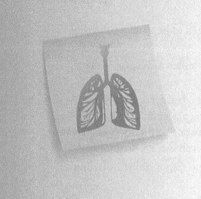

10 Integrative physiology

The purpose of this chapter is to examine some areas that affect the systems discussed in this textbook. The aim is not to provide an extensive overview of the topics in question but to demonstrate the integrative nature of physiological systems and the effect that certain applied examples have on these systems. The examples discussed are the effects of acute and chronic aerobic exercise on the cardiovascular system, the effect of nutrition on health, with particular focus on obesity, and the effect of ambient temperature and pressure on physiological systems.

Key concepts

- The functioning of physiological systems is affected by numerous external factors.
- Acute aerobic exercise causes an increase in cardiac output, which is due to increases in stroke volume and heart rate.
- Acute aerobic exercise leads to increases in mean arterial pressure due to increases in cardiac output.
- A period of aerobic training results in numerous adaptations to cardiovascular system function.
- The prevalence of obesity has significantly increased in recent times and numerous health consequences are associated with this disorder.
- The regulation of appetite involves psychological and physiological interactions.

continued

- There are four main mechanisms of heat transfer between the body and the environment.
- As altitude increases, partial pressure of oxygen decreases, resulting in both beneficial and deleterious effects on a variety of systems.

10.1 EXERCISE AND THE CARDIOVASCULAR SYSTEM

Exercise causes both acute and chronic changes in physiological function of almost all body systems. One of the most pronounced effects is on the cardiovascular system in order to supply oxygen and nutrients to muscle and to remove metabolic waste products.

Acute effects of exercise on cardiac output

- Following an increase in metabolic activity in skeletal muscle, cardiac output must increase in order to supply oxygen and to remove carbon dioxide and metabolic waste products.
- At rest, cardiac output is approximately 5L/min, but during exercise this can increase to 35–40L/min depending on how trained the individual is.
- As covered in Chapter 5, cardiac output is the product of heart rate and stroke volume. The increase in cardiac output observed during exercise is the result of increases in both heart rate and stroke volume.
- The increase in heart rate is due largely to changes in central nervous system activation. Following the start of exercise, vagal tone is removed from the sinoatrial node resulting in an increased frequency of pacemaker potentials. When vagal tone is completely removed, sympathetic tone is increased leading to a further increase in the frequency of pacemaker potentials and heart rate. In addition, circulating hormones such as adrenaline and noradrenaline also produce an increase in heart rate up to an age-related maximum.
- The increase in stroke volume is due to the effect of preload. During exercise, venous return is increased as a result of a combination of an increase in sympathetic activity to veins and the use of skeletal muscle and respiratory muscle pumps. This increase in venous return leads to an increase in end-diastolic volume and an increase in cardiac muscle fibre stretch. This causes an increase in stroke volume.
- As will be covered in the next section, exercise tends to increase blood pressure and will consequently increase afterload. This would normally reduce stroke volume. However, the effect of preload is greater than that of afterload meaning that stroke volume is increased.
- **Dehydration** is a common occurrence during exercise as sweat losses tend to exceed fluid intake. Large sweat losses lead to a decrease in plasma volume and a consequent increase in blood viscosity. In turn, this leads to a reduced venous return, a reduced preload, and a reduction in stroke volume. In order to maintain cardiac output, heart rate must increase – an effect termed cardiovascular drift.

Exercise and the cardiovascular system

Acute effects of exercise on blood pressure

- Chapter 5 covered the mechanisms that regulate mean arterial pressure. An understanding of these mechanisms assists with explaining the short-term effects of different types of exercise on mean arterial pressure. Mean arterial pressure is the product of cardiac output and total peripheral resistance.
- Resistance exercise can lead to an increase in systolic blood pressure of up to 240mmHg, and an increase in diastolic pressure is also observed.
- Resistance exercise results in an increase in sympathetic tone. In turn, this causes an increase in cardiac output, which is one explanation for the increase in mean arterial pressure observed.
- In addition, physical compression of vessels as a result of repeated muscular contraction causes an increase in total peripheral resistance.
- The amount of muscle mass activated to complete the exercise has a major effect on the extent of the increase in blood pressure observed. Use of a greater volume of muscle mass results in a greater increase in blood pressure.
- During continuous steady state aerobic exercise, systolic blood pressure may increase to approximately 150mmHg with little change in diastolic pressure. A gradual decline from the peak systolic pressure is observed as exercise continues.
- Increased systolic blood pressure is due to the increase in cardiac output that accompanies exercise. Total peripheral resistance is reduced as a result of an increase in vasodilation. This explains the gradual decline in systolic pressure as exercise continues.
- A gradual increase in intensity throughout the period of exercise results in continued increases in systolic blood pressure up to approximately 200mmHg. As exercise intensity increases, cardiac output increases in a linear fashion, which, in turn, increases systolic blood pressure. Diastolic blood pressure tends to be unaffected by this type of exercise.

Chronic effects of aerobic exercise on the cardiovascular system

- A sustained period of aerobic exercise training has many beneficial effects on the cardiovascular system. They all lead to increased efficiency of the system.
- One of the main benefits is an increase in left ventricular volume, which may be up to 25% larger in trained individuals than in non-trained individuals.
- Aerobic exercise training results in an increase in both calcium sensitivity and the number of myocardial sarcomeres, which ultimately leads to an increase in stroke volume. Consequently, trained individuals have a greater resting and exercising stroke volume than untrained individuals.
- The increase in left ventricular volume that accompanies aerobic exercise training is not permanent. It returns to pre-training volume after the cessation of training programme.
- A single exercise session leads to an increase in plasma volume, and a period of aerobic exercise training substantially increases plasma volume. This is largely due

to an increase in the retention of albumin and the consequent retention of water that is necessary to maintain plasma osmolality. This increase in plasma volume may be one explanation for the increase in left ventricular volume.

- As resting and during submaximal exercising stroke volume is increased due to the increase in left ventricular volume.
- Heart rate at rest and during submaximal exercise is lower in trained individuals than in untrained individuals. At rest, this is largely due to an increase in vagal tone that leads to reduced pacemaker potentials in the sinoatrial node.
- Trained individuals exhibit an increase in maximal cardiac output. This is primarily due to the increased stroke volume that accompanies aerobic exercise training.
- A period of aerobic exercise training results in an increased efficiency of blood distribution and extraction of oxygen at the skeletal muscle. In addition, more mitochondria are present within the skeletal muscle and the activity of oxidative enzymes is increased, leading to more efficient oxidative metabolism.
- A period of aerobic exercise training results in an increase in activity of enzymes involved in aerobic metabolism, as well as an increase in the number of mitochondria resulting in an increased capacity for oxidative metabolism.
- The ability to extract oxygen is increased following aerobic exercise training.

10.2 NUTRITION AND HEALTH

The role of nutrition in health and disease has been the subject of much recent research. The rising prevalence of obesity is of particular concern, given the numerous health consequences that can arise and the cost that this places on health services.

- **Obesity** can be defined as an excessive accumulation of fat which results in adverse health problems.
- While obesity is not a new disorder, its prevalence has increased substantially over the last 50 years. This is particularly evident in Western societies where up to 25% of the adult population in some countries are classed as being clinically obese. Obesity is more prevalent in females than in males.
- The prevalence of childhood obesity has also increased substantially in recent years.
- An increase in fat mass is the result of an imbalance in energy intake and energy output. When energy intake exceeds energy output, excess energy is stored in the form of fat. The reasons for this imbalance are complicated and subject to much debate.
- A number of factors affect energy intake. These include physiological factors such as the homeostatic regulation of appetite as well as psychological and behavioural factors such as habit or social situation.
- Energy output consists of **basal metabolic rate**, **dietary-induced thermogenesis**, and voluntary physical activity.
- The trend of increasing prevalence of obesity may be due to changes in diet and/ or voluntary physical activity over time.

Nutrition and health

- In the UK, there has been a trend of an overall reduction in energy intake over the last 40 years, with the proportion of intake from fat and carbohydrate increasing and decreasing, respectively, during that period. This suggests that a reduction in physical activity patterns may be of particular importance when considering the reasons for the increased prevalence of obesity.
- Approximately 40% of adult males and 30% of adult females achieve government recommended guidelines for weekly participation in physical activity (a total of 150 minutes of moderate intensity aerobic activity per week). The proportion of adults classed as inactive has decreased recently, suggesting that strategies to improve participation in physical activity have had some effect.

Measurement of fat mass

- The ability to obtain accurate measurements of body composition, and fat mass in particular, is important when attempting to determine whether an individual is obese.
- **Body mass index (BMI)** is a commonly used measurement of body composition and is calculated by dividing body mass (kg) by height (m) squared. The calculation of BMI assumes that all deviation in body mass is due to height, and therefore does not consider lean tissue, which causes problems with its use in certain populations.
- A BMI of $18–24.9 \mathrm{kg/m^2}$ is considered to be ideal, with a BMI of $25–29.9 \mathrm{kg/m^2}$ being overweight and $>30 \mathrm{kg/m^2}$ being obese.
- Abdominal obesity, where fat is stored predominantly around the abdomen, is of particular interest, as studies have suggested that individuals with a high ratio of waist to hip circumference (an indicator of abdominal obesity) are more likely to suffer from cardiovascular disease. This is likely due to storage of fat around a number of important organs in the abdominal area. Consequently, the measurement of waist circumference is encouraged in clinical settings in addition to the measurement of BMI.
- Numerous methods, each with relative advantages/disadvantages related to accuracy and cost, are available to measure, or calculate, fat mass. Body density can be measured using the underwater weighing technique. It is calculated by subtracting the individual's weight in water from their weight in air. Body fat is then calculated from body density. Although this technique is relatively accurate, it requires specialist equipment and can be very costly.
- Fat mass can be calculated by subtracting fat-free mass from total body mass. Approximately 73% of all fat-free mass is water; therefore determination of total body water, via the use of stable isotopes, can be used to determine fat mass.
- **Dual energy X-ray absorptiometry** was initially developed in order to examine bone density, but can also be used to measure fat and lean mass.
- **Bioelectrical impedance analysis** is a commonly used method of fat mass determination which involves passing a small current through the body. As fat is a poor conductor of electricity, the resistance to the current increases with increasing fat mass. This impedance value can then be used to calculate fat mass. This method is commonly used and relatively cheap, but its accuracy has been questioned.

Regulation of appetite

- The feeling of hunger is a complicated area; it encompasses both physiological and psychological factors and is a key determinant of energy intake.
- The physiological regulation of appetite involves a complex interaction between the stomach, intestines, pancreas, adipose tissue, and the hypothalamus.
- **Satiation** can be defined as the processes that result in an individual stopping feeding whereas **satiety** is the prolonged feeling of fullness following termination of eating. Much of the physiological research in this area has focused on satiety, as increasing the satiety period (i.e. the time between snacks/meals) may lead to a reduction in overall daily energy intake.
- The stomach is involved in the satiety process as a result of gastric distension. This occurs as the stomach fills, and feedback from the vagus nerve to the hypothalamus assists in the feeling of fullness after termination of food intake.
- **Anorexigenic** pathways within the hypothalamus involve neurons that secrete **pro-opiomelanocortin (POMC)**. The secretion of POMC leads to an increase in alpha-melanocyte stimulating hormone (alpha-MSH), which acts on melanocortin 3 and 4 receptors and results in a reduction in energy intake.
- **Orexigenic** pathways within the hypothalamus involve neurons that secrete **neuropeptide Y**, which acts on Y1 and Y5 receptors, resulting in an increase in food intake. In addition, activation of these neurons leads to inhibition of anorexigenic pathways.
- A number of hormones secreted from the gastrointestinal tract have been shown to influence activation of these pathways. Ghrelin is the only one of these that appears to increase subjective feelings of hunger. Ghrelin is a hormone secreted from the fundus of the stomach and leads to activation of orexigenic pathways. Secretion of ghrelin is reduced after meal ingestion and steadily rises until the next meal, suggesting a role for this hormone in meal initiation. Despite this, a consistent negative correlation with BMI has been observed.
- Peptide YY (PYY), **glucagon like peptide-1 (GLP-1)**, and **pancreatic polypeptide (PP)** are secreted from the intestines (PYY and GLP-1) and the pancreas, and have all been shown to reduce food intake when administered peripherally. This is likely to be due to inhibition of orexigenic pathways.
- **Leptin** is a hormone that is released from adipose tissue and is strongly involved in long-term energy balance. Secretion of leptin is related to insulin release.

Genetics and obesity

- Much recent research has focussed on the role that genes play in the development of obesity. In relation to obesity, genetic variation may affect the regulation of appetite, the process of fat breakdown, and/or the effectiveness of physical activity as a weight loss strategy.
- Twin studies suggest that to some extent there is a genetic basis for obesity. However, there is a significant effect of environmental factors.

- The genetic basis of obesity can include two main areas: **monogenic syndromes** and **susceptibility genes**.
- Monogenic syndromes are extremely rare and are the result of a defect in a single gene. These syndromes often result in a very severe form of obesity and have extremely high health costs and a likelihood of related health consequences.
- Leptin receptor deficiency is one example of a monogenic syndrome. This disorder involves a defect in the gene that encodes the leptin receptor meaning that, although the hormone may be secreted, leptin cannot exert its effect. Consequently, individuals constantly feel hungry even after very large caloric intakes.
- Numerous susceptibility genes have been examined with respect to their role in the development of obesity. The majority of these are related to the regulation of appetite (such as gene variants encoding POMC and melanocortin receptors) or fat oxidation. While numerous studies have reported increased BMI and/or fat mass in those with certain gene variants, the importance of these genes in the development of obesity is still a matter of serious debate.

Health consequences of obesity

- Obesity may result in a variety of comorbidities. Obese individuals have a 25–50% increase in likelihood of premature death when compared with non-obese individuals.
- The likelihood of obese individuals developing hypertension is up to 400% more than for non-obese individuals. Increases in renal tubular reabsorption and circulating blood volume as a result of activation of the renin–angiotensin–aldosterone system are thought to be important mechanisms behind the increase in blood pressure associated with obesity. The increase in blood volume, in particular, will lead to an increase in left ventricular volume.
- Obese individuals often exhibit a degree of **hyperlipidaemia**, i.e. an abnormally high concentration of triglycerides in the blood. The role of low and high density lipoproteins in fat transport is discussed in Chapter 8.
- Obesity tends to result in an increase in circulating triglycerides and low density lipoproteins as well as a reduction in circulating high density lipoproteins. This greatly increases the risk of atherosclerosis.
- Atherosclerosis is the build-up of fatty plaques within blood vessels. This process begins early in life and the plaques are built up over time. A number of factors may affect the speed at which atherosclerosis occurs, with dyslipidaemia being a major determinant.
- Atherosclerosis leads to narrowing and reduced compliance of the blood vessel. This results in an increase in total peripheral resistance and partly explains the increased risk of hypertension in obese individuals.
- As outlined in Chapter 5, hypertension can lead to an increased risk of myocardial infarction, stroke, and renal disease.
- A close relationship between BMI and the risk of development of type 2 diabetes mellitus has been demonstrated. An increase in fat mass leads to insulin resistance

as a result of an inability to activate insulin receptors. There is a reduced ability to uptake glucose into the target tissue and this leads to hyperglycaemia.

- In addition to the hyperglycaemia that occurs in type 2 diabetes mellitus, the elevated blood insulin leads to an increase in circulating triglyceride concentrations and the potential consequences of dyslipidaemia.

- **Osteoarthritis** is a common complication associated with obesity; there is no compensatory increase in the size of the skeleton as fat mass increases. This puts more pressure on certain joints, such as the knees, which, in turn, leads to mobility issues.

- **Sleep apnoea** is a common side effect of obesity. The presence of fat in the neck and chest area reduces the compliance of the lungs as well as total lung volume. This results in a change in breathing patterns and the volume of air inhaled and exhaled in each breath. This may lead to the use of respiratory equipment at night to assist with breathing.

- As outlined in Chapter 8, the purpose of the gall bladder is to store bile prior to its release into the small intestine. If excess cholesterol is present, this can lead to the production of gallstones. Therefore there is a close association between obesity and the risk of **cholelithiasis.**

- The risk of breast and uterine cancer in obese females is greater than in non-obese women. This may be due to the production of oestrogen by excess adipose tissue.

- The risk of colon, pancreatic, and renal cancer is greater in obese individuals than in non-obese individuals. This is likely to be due to the increased circulating insulin concentrations observed with obesity.

- Obese females are at risk of amenorrhoea and eumenorrhoea. This is likely to be the result of an increase in testosterone secretion brought about by elevated circulating insulin concentration.

- Obese males are at risk of reduced sperm count, which, again, may be due to changes in circulating sex hormone concentration.

- In addition to the many physiological effects of obesity, it is important to remember the psychological effects of obesity, which have a detrimental effect on quality of life. These include, but are not limited to, reduced self-esteem and increased incidence of depression.

10.3 THE EFFECTS OF ALTITUDE

The physical environment

- There are significant differences between the environments at altitude and at sea level.
- An increase in altitude is associated with:
  - a decrease in temperature
  - increased amounts of solar and other ionizing radiation
  - a reduction in barometric pressure—it is this aspect which will be considered in further detail.

The effects of altitude

- Atmospheric pressure at sea level is equivalent to 760mmHg. At a height of about 8,000 feet, barometric pressure is just over 500mmHg. At the height of Mount Everest (29,029 feet), it is about 240mmHg, and at 35,000 feet it is about 150mmHg.
- Although the barometric pressure decreases with increasing altitude, the composition of the air in the environment does not change.
- The consequence of this is that the partial pressure of oxygen in the environment decreases.
- Assuming that oxygen contributes 20.9% of atmospheric air, po_2 at sea level is $(760 \times 20.9)/100 \approx 160$mmHg. At an altitude of about 16,000 feet (the highest recorded altitude of human habitation), atmospheric pressure is about 375mmHg, and so po_2 is $(375 \times 20.9)/100 \approx 60$mmHg.
- Remember from Chapter 6 that gas movements occur down their partial pressure gradients.
- This means that since there is 'less' oxygen in the atmosphere, there will be reduced amounts in the blood.
- Lack of oxygen is called **hypoxia**.
- Hypoxia may produce acute effects (e.g. in people taking holidays at altitude) or chronic effects (e.g. in those living and working at altitude).

Acute effects of hypoxia

- The initial acute response to hypoxia is hyperventilation—an increase in both the rate and depth of breathing.
- Since the environmental po_2 decreases, so does the po_2 of plasma, although this has to be reduced to about 60mmHg before ventilation is stimulated (normal levels are about 100mmHg).
- Reductions in plasma po_2 are detected by the carotid and aortic bodies, as described in Chapter 6, and this drives ventilation.
- As a consequence of this hyperventilation, excessive amounts of CO_2 are blown off. This subsequently reduces the drive to ventilate originating from the central chemoreceptors. It also induces a respiratory alkalosis.
- Cardiovascular changes include increased heart rate and diversion of blood to essential circulations (e.g. cerebral) from less essential circulations (e.g. skin).
- The reduction in plasma CO_2 levels occurring as a consequence of hyperventilation pushes the O_2–Hb dissociation curve to the left.
- This means that for a given partial pressure of oxygen, Hb is more saturated.
- A number of potential health issues are related to acute hypoxia—this is called mountain sickness.
- Symptoms may include nausea, vomiting, and severe headache. These will disappear with acclimatisation.
- Vasoconstriction of blood vessels in the pulmonary circulation may occur, resulting in the formation of pulmonary oedema, which may compromise breathing.

Chronic effects of hypoxia

- Humans living and working at altitude quickly acclimatize to the different environment they find themselves in. This can start to occur as soon as 24 hours after ascent.
- One of the first responses is a resetting of central chemoreceptors.
- The respiratory alkalosis that occurs as a consequence of hyperventilation is compensated by the increased renal excretion of bicarbonate ions. This may take up to seven days to occur.
- Similar changes occur in the cerebrospinal fluid. This means the central chemoreceptors are reset to the new lower level of arterial CO_2.
- Over the initial few weeks at altitude, there is an increase in both the number of red blood cells and the quantity of Hb in each cell.
- The renal hormone erythropoietin is responsible for the increased number of red blood cells—its release is stimulated by hypoxia.
- There is also an increase in the amount of 2,3-DPG in the cells. This produces a Bohr effect and increases oxygen delivery to tissues.
- Alongside these responses of the vascular system, there is also a maintained increase in cardiac output, which is achieved by increasing heart rate. Together, these responses ensure that oxygen delivery to tissues is maintained in the face of a reduced po_2 in the environment.

Looking for extra marks?

Erythropoietin is a hormone released primarily from the kidney when hypoxic conditions are encountered. It is not clear how changes in oxygen levels are detected, but one possibility is that haem-like proteins (similar to those found in Hb), when exposed to hypoxic conditions, promote the activity of the gene responsible for producing erythropoietin. Erythropoietin is a banned substance in athletics as it potentially gives competitors an advantage in terms of their oxygen-carrying capacity, which has an impact on athletic performance. However, some athletes do train at altitude prior to competition at lower levels.

10.4 THERMOREGULATION

Basic principles of thermoregulation

- The ability to maintain an appropriate body temperature is an absolute requirement for survival. Disruptions to body temperature come from environmental challenges (e.g. weather changes) and self-induced challenges (e.g. exercise). The process of regulating body temperature is known as thermoregulation.
- It is possible to distinguish between core body temperature (temperature of brain, thorax, and abdomen) and shell temperature (temperature of the skin). Whilst the latter may vary, the former is closely regulated between 36.5 and 37°C.

Thermoregulation

- It is essential that temperature remains fairly constant so that enzymes within the body work at their optimum rate. At high temperatures there is a risk that the enzymes will become denatured and fail to work.
- The lower and upper limits of the core temperature are 33°C and 42°C respectively—beyond these death occurs and even at these temperatures there is significant disruption to normal physiology. Generally, the body deals with an increased temperature better than a decreased temperature
- In order to maintain a constant body temperature heat losses must be balanced by heat gains. Heat transfers between the body and environment by four mechanisms which may be influenced by clothing, sweating etc:
  - conduction
  - convection
  - evaporation
  - radiation.
- The central control of thermoregulation is maintained by the activity of neurons in the hypothalamus.
- The hypothalamus contains the set point for temperature, i.e. it intrinsically 'knows' that body temperature should be 37°C.
- Both peripheral and central thermoreceptors provide an afferent input to the hypothalamus—these are non-myelinated or small-diameter afferent neurons.

Looking for extra marks?

Both the hypothalamus and the skin contain thermoreceptors, which respond to either an increase or a decrease in temperature—warm and cold receptors respectively. There are believed to be more cold receptors than warm receptors. Thermoreceptors display phasic adaptation, i.e. once the temperature is stable they switch off. An everyday example of this is stepping into a hot shower—the water may initially feel very hot, but this quickly subsides. The temperature of the water has not changed; rather, the thermoreceptors have stopped signalling.

Responses to increased temperatures

- An increase in body temperature is resolved by increasing heat loss and decreasing heat gain and production.
- Responses to achieve this may be both physiological and behavioural (e.g. seeking shade, dressing appropriately).
- A reduction in vasoconstrictor responses to blood vessels in the skin increases blood flow. As more blood flows through the skin, so more heat is lost to the environment.
- Heat loss can be increased further by sweating—evaporation of sweat from the skin produces cooling. This may be reduced when the relative humidity is high, therefore reducing its cooling effect.
- Sweat is produced by **eccrine glands** located in the skin, with the majority being on the trunk.

- Appreciable amounts of sodium chloride may be lost in sweat. Thus thermoregulatory responses may have impacts on other physiological systems, which is an example of how systems do not work in isolation from each other.
- The autonomic nervous system provides some control over sweating. Anatomically sympathetic fibres, which, unlike most other sympathetic fibres, release acetylcholine and increase its production. Excessive sweating (hyperhidrosis) can be treated by injections of Botox, which interferes with autonomic activity.

Responses to decreased temperature

- A decrease in body temperature is resolved by decreasing heat loss and increasing heat gain.
- In contrast to an increased body temperature, a decreased body temperature results in increased vasoconstrictor impulses to blood vessels in the skin. This reduces skin blood flow and therefore heat loss to the environment.
- Heat production can be increased by **shivering**.
- Shivering is an asynchronous contraction of skeletal muscle. It produces no purposeful movement, but heat is generated as a by-product of contraction.
- Such a response cannot be maintained for asustained period of time.
- Heat can also be produced by a variety of processes collectively known as **non-shivering thermogenesis**.
- Essentially, non-shivering thermogenesis represents increased metabolism in brown adipose tissue. Metabolism may be stimulated by a variety of hormones including adrenaline and the thyroid hormones.

Looking for extra marks?

It is possible to reset the hypothalamic set point of 37°C. A good example of this is the temporary change seen in thermoregulation in response to infections. In defending the body against infections the immune system may produce pyrogens, which temporarily reset the set point to a higher value. The increase in body temperature may help defend the body against the infection. However, this process cannot go unchecked since prolonged increased body temperature is associated with disorders of the nervous system (e.g. febrile convulsions).

Check your understanding

Why is heavy resistance exercise unlikely to be prescribed for an individual suffering from hypertension? (*Hint: Consider the effects of resistance exercise on mean arterial pressure*)

Discuss the effects of the administration of glucagon-like peptide 1 and peptide YY on energy intake. (*Hint: Consider the effects of these hormones on orexigenic and anorexigenic pathways*)

Discuss the response to a reduction in core body temperature. (*Hint: Consider the mechanisms which increase heat generation and reduce heat loss*)

Glossary

2,3-DPG a metabolite found inside red blood cells which decreases the affinity of oxygen in binding to haemoglobin

17-β-oestradiol the principal oestrogen found in females

acetylcholinesterase enzyme that cleaves acetylcholine from nicotinic receptors on the motor endplate of a skeletal muscle fibre

acidosis reduction in blood pH below the normal range

acrosome the head region of a sperm cell which contains hydrolytic enzymes

actin key protein found in muscle fibres

action potential (AP) a transient reversal of the resting membrane potential characterized by the inside of the neuron becoming positively charged

active hyperaemia an increase in metabolites that leads to vasodilation of arterioles and an increase in blood flow to active tissue

adenylate cyclase enzyme responsible for the production of cAMP from ATP

adrenal glands paired glands lying above the kidney divided into the adrenal cortex, itself divided into three layers, and the adrenal medulla

adrenaline hormone released from the adrenal medulla when the sympathetic nervous system is activated producing a range of physiological responses (e.g. increased heart rate)

adrenocorticotrophic hormone (ACTH) hormone released from the anterior pituitary gland which stimulates the adrenal cortex

afferent neuron a neuron which relays action potentials towards the central nervous system

after-hyperpolarization the final stage of the action potential when there is a drop in membrane potential below the usual value of the resting membrane potential; hyperpolarization is the general term for the inside of neuron becoming more negatively charged

afterload the effect that pressure within the aorta has on stroke volume

albumin main protein found in blood plasma

aldosterone hormone that is part of the renin–angiotensin–aldosterone system which acts on the distal tubule and collecting ducts to increase sodium and water reabsorption

alkalosis increase in blood pH above the normal range

alveolar ducts conducting tubes which give rise to the alveolar sacs

alveoli site of gas exchange in the airways

ampulla of Vater area where the pancreatic duct and the common bile duct join

angina pectoris localized pain due to reduction in oxygen delivery to the myocardium

angiogenesis growth of new capillaries

anterior towards the front of the body

anterior pituitary gland region of the pituitary gland which secretes hormones and which itself is controlled by hormones released from the hypothalamus

antidiuretic hormone (ADH) see vasopressin

anorexigenic neural pathways that lead to a reduction in energy intake and/or an increase in energy expenditure

anus the external opening of the rectum responsible for the elimination of faeces

aortic bodies chemoreceptors responsible for measuring oxygen levels in arterial blood

aortic valve valve between the left ventricle and the aorta

apical membrane membrane of a cell closest to the lumen

apneustic centre neurons which prolong inspiration by preventing the inhibition of inspiratory neurons

apolipoprotein lipid-binding proteins involved in the transport of fats

aquaporins membrane channels which, when inserted, lead to water reabsorption in the collecting ducts of the kidney

arachnoid mater the middle layer of the meninges

arginine vasopressin see vasopressin

arteries relatively large blood vessels that carry blood away from the heart

arterioles continuations of arteries that assist in carrying blood away from the heart and determine blood flow to tissues

astrocyte type of glial cell which contributes to the formation of the blood–brain barrier

atherosclerosis build-up of plaques within blood vessels that may lead to reduced blood flow to tissues

atresia the death of all follicles, bar one, which occurs each month during a normal ovarian cycle

atria chambers of the heart that receive deoxygenated blood from the systemic circulation (right atrium) or oxygenated blood from the pulmonary circulation (left atrium)

atrial natriuretic peptide hormone secreted from the right atrium in response to an increase in blood pressure which leads to water and sodium secretion

atrioventricular node bundle of cells in the lower right atrium that propagate action potentials into the ventricles

autocrine control endocrine control characterized by chemical substances influencing the activity of cells from which they are released

autonomic nervous system (ANS) efferent part of the peripheral nervous system which regulates the activity of smooth muscle, cardiac muscle, and glandular secretions

autoregulation change in pressure within an arteriole that leads to a change in circulating levels of metabolic by-products and a change in blood flow to tissue

basal ganglia a functional region of the brain involved in the regulation of movement

basal metabolic rate minimum energy requirements in a thermally neutral non-active state

basolateral membrane membrane of a cell furthest from the lumen

bicarbonate ion (HCO$_3^-$) the principal form in which CO_2 is transported from tissue to the lungs

bicuspid valve valve between left atria and ventricle

bile liquid produced within the liver and stored in the gall bladder which aids in the digestion of nutrients

bioelectrical impedance analysis measurement tool used to determine body composition

bladder part of the renal system that is involved in storing urine prior to elimination via the urethra

blastocyst hollow ball of cells found about 14 days after fertilization; an inner mass develops into the embryo whilst the outer layer contributes to the formation of the placenta

blood–brain barrier structure (formed from astrocytes and the specialized capillaries of the cerebral circulation) which ensures that there is no unregulated movement of substances from blood plasma to neural tissue

blood pressure pressure exerted on blood vessel walls by the blood

Glossary

body (stomach) central region of the stomach

body mass index (BMI) common calculation based on height and body mass that is often used to determine of body composition

Bohr effect rightward shift of the oxygen–haemoglobin saturation curve caused by increased levels of CO_2, decreased pH, or increased temperature.

Bowman's capsule part of the renal corpuscle involved in storing filtrate from the glomerulus

brainstem lowest portion of the brain, formed from the midbrain, pons, and medulla and continuous with the spinal cord

bronchioles conducting tubes of the airways that lack cartilage

brush border membrane layer of epithelial cells of villi nearest the lumen of the digestive system

buffer a substance or solution which temporarily regulates blood pH by the addition or removal of hydrogen ions

bundles of His specialized cells within the heart that propagate action potentials from the atrioventricular node towards the myocardium

calcitonin a thyroid hormone which inhibits osteoclasts

calmodulin protein found in smooth muscle fibres which, when bound to calcium, leads to activation of myosin light chain kinase

capacitation a calcium-mediated process which sperm undergo in the female reproductive tract to ensure that they are capable of fertilizing an ovum

capillaries blood vessels that surround tissue and are the site of gas exchange

carbaminohaemoglobin haemoglobin with CO_2 bound to it

carbohydrates a range of molecules which are the primary energy source of cells

carbonic acid–bicarbonate buffer system system that increases or decreases hydrogen ion concentration as a result of changes in carbon dioxide production

carboxyhaemoglobin haemoglobin that is carrying carbon dioxide

cardiac cycle sequence of events that occurs during one heart beat and consists of systole and diastole

cardiac muscle fibre mononucleated cell found in cardiac muscle

cardiac output volume of blood ejected from the heart every minute

cardiac region region of the stomach near the lower oesophageal sphincter

cardiac tamponade filling of the pericardium with fluid leading to increase in pressure on the heart

carotid bodies chemoreceptors responsible for measuring oxygen levels in arterial blood

caecum first section of the large intestine

central chemoreceptors chemoreceptors on the surface of the medulla which monitor hydrogen ion concentration in CSF and therefore indirectly CO_2 levels in arterial blood

cerebellum region of brain concerned with the control and execution of movement

cerebral cortex the outermost layer of the cerebrum concerned with the highest levels of analysis of sensory information, abstract thought, and the planning of movement

cerebrospinal fluid (CSF) fluid formed in the lateral ventricles of the brain which flows over the brain and spinal cord in the subarachnoid space

chemoreceptors receptors which monitor and respond to changes in the levels of chemicals (e.g. the amount of glucose in the plasma)

chief cells cells within the stomach that secrete pepsinogen

chloride shift the exchange of bicarbonate ions in red blood cells for chloride ions present in the plasma

cholecystokinin (CCK) hormone that regulates the activity of the sphincter of Oddi

cholelithiasis gallstones in the gall bladder, usually as a result of the presence of excess cholesterol

chordae tendinae tendons that connect papillary muscles to heart valves and the heart wall

chylomicron lipoprotein particle containing triglycerides, phospholipids, and cholesterol

chyme partly digested nutrients in the stomach following emulsification and other chemical processes

circadian rhythms cyclical 24 hour changes seen in a number of physiological processes (e.g. body temperature, levels of certain hormones)

collecting duct section of a renal tubule before entering the ureter and the main site of action of arginine vasopressin

colon second section of the large intestine consisting of the ascending, transverse, descending, and sigmoidal sections

common bile duct duct connecting the gall bladder to the small intestine which transports bile

common hepatic duct duct connecting the liver to the common bile duct which transports bile directly to the intestine rather than for storage in the gall bladder

comparator general term for a structure which compares two pieces of physiological information (e.g. actual body temperature versus desired body temperature)

compliance a measure of stretch associated with blood vessels

concentric contraction isotonic muscular contraction that results in an increase in force with a reduction in muscle length

conduction system system in the heart consisting of the sinoatrial node, atrioventricular node, bundles of His, and Purkinje fibres which leads to contraction of cardiac muscle fibres

connective tissue one of four fundamental tissue types responsible for the support and linkage of cells

coronary arteries blood vessels that supply the myocardium with oxygenated blood

corpus luteum the empty follicle left behind after ovulation has occurred

corticotrophin releasing hormone (CRH) hypothalamic hormone which stimulates the release of ACTH

cortisol hormone released from the adrenal cortex which has anti-inflammatory effects and is essential in responding to stress

cross-bridge link between the myosin head and the actin-binding site that is essential for muscular contraction

cross-bridge cycling four-stage process that leads to muscular contraction

crypt of Lieberkuhn groups of secretory cells at the end of each villus in the digestive system

cyclic AMP (cAMP) a second messenger formed from ATP which activates intracellular protein kinases

cystic duct duct joining the liver to the gall bladder which transports bile for storage

cytoskeleton the internal scaffolding, formed of proteins, which gives a cell its 3D shape

dehydration condition resulting from large volumes of water loss

dendrites extensively branched projections arising from the cell body of a neuron and responsible for the collection of incoming action potentials to the neuron

dense bodies structural protein of smooth muscle fibres

depolarization change in membrane potential such that the inside of a neuron becomes less positively charged (e.g. as occurs during the initial phase of an action potential)

detectors general term for structures which gather information about either the internal or external environment

Glossary

diaphragm a sheet of muscle separating the abdominal and thoracic cavities; contraction of this muscle is primarily responsible for inspiration

diastole filling phase of the cardiac cycle

diastolic pressure the pressure within systemic circulation arteries at the end of diastole

dietary-induced thermogenesis energy expended as a result of nutrient digestion and absorption

digestion physical and chemical processes associated with the breakdown of nutrients in preparation for absorption

disaccharide two monosaccharides bound together following a dehydration synthesis reaction

distal tubule section of a renal tubule before the collecting duct in which some tubular reabsorption occurs

dorsal respiratory group region of the respiratory centre responsible for inspiration

dorsal root ganglia cell bodies of sensory neurons whose dendrites are located in the periphery and whose axon terminals terminate in the spinal cord

dual energy X-ray absorptiometry measurement tool used to determine body composition

duct of Wirsung duct connecting the pancreas to the small intestine

duodenum first section of the small intestine

dura mater the outermost layer of the meninges

eccentric contraction isotonic muscular contraction which results in an increase in force with an increase in muscle length

eccrine gland a tubed sweat gland

effectors general term for structures which generate a response in a physiological system (e.g. the contraction of muscle)

efferent neuron a neuron which relays action potentials away from the central nervous system

ejaculatory duct the region formed where the vas deferens merges with the outflow from the seminal vesicle

electrocardiogram representation of the electrical activity of the heart that leads to muscular contraction

end-diastolic volume the volume of blood in the ventricles at the end of diastole

endocardium inner layer of the heart wall consisting of connective tissue and a layer of endothelium

endocrine control release of chemical substances into the plasma which influence the activity of cells (target cells) some distance away from the site of release

endometrium the lining of the uterus

endomysium thin layer of connective tissue surrounding a skeletal muscle fibre

endoplasmic reticulum a widespread organelle associated with the production of proteins

enteric nervous system part of the autonomic nervous system that is specific to the digestive system and responds to changes in the environment

ependyma glial cells which line the lateral ventricles of the brain and are involved in the formation of cerebrospinal fluid

epicardium layer of connective tissue on the outside of the heart wall

epididymis a tubular structure in the testes into which seminiferous tubules drain

epimysium layer of connective tissue surrounding skeletal muscle

epithelial tissue one of four fundamental tissue types which is responsible for covering either internal or external structures

erythrocytes red blood cells produced as a result of erythropoiesis in active bone marrow

erythropoietin hormone secreted from the peritubular capillaries in response to a reduction in oxygen delivery to the kidney which leads to the formation of erythrocytes

excitation–contraction coupling process of increasing cytosolic calcium concentration in muscle fibres which leads to muscular contraction

expiratory reserve volume (ERV): the volume of air that can be exhaled over and above a normal tidal volume

external elastic membrane layer of elastic fibres found in the tunica media layer of arteries

extracellular fluid (ECF) fluid found outside cells (e.g. plasma) which is characterized by a large sodium concentration

Fallopian tubes arm-like 'extensions' of the uterus, the distal portion of which is the site of fertilization of the ova by sperm

fascicles bundles of skeletal muscle fibres surrounded by the perimysium

feedforward anticipatory behaviour which aims to limit disruption to physiological systems and therefore minimize any need for negative feedback control

fertilization the union of male and female gametes—sperm and egg

follicle stimulating hormone (FSH) hormone released from the anterior pituitary gland involved in the regulation of the reproductive system

force–velocity curve relationship that describes the effect of speed of muscular contraction on force production

fundus region of the stomach that is largely responsible for increasing in size with little increase in intragastric pressure

gall bladder ancillary organ of the gastrointestinal system responsible for storage of bile prior to secretion into the small intestine

gap junctions specialized connections between cells that allow movement of ions

gastrin hormone secreted from G-cells that leads to hydrochloric acid secretion

gastric emptying movement of fluid or chyme from the stomach to the small intestine

gastrulation a process in embryo formation which results in the formation of germ cell layers

G-cells cells within the stomach that secrete gastrin

ghrelin hormone secreted from cells in the fundus that leads to increased secretion of growth hormone and has been implicated in the regulation of appetite

glial cells cells in the nervous system whose overall role is to nuture neurons

glomerular filtrate liquid produced following filtration of plasma in the glomerulus that is passed into the renal tubule

glomerulus bundle of capillaries within a renal corpuscle and the site of filtration of plasma

glucagon a pancreatic hormone which produces an increase in plasma glucose levels

glucagon-like peptide 1 incretin hormone involved in insulin secretion and implicated in the regulation of appetite

glucose transporter 2 (GLT2) intestinal transporter responsible for transport of glucose across the basolateral membrane of enterocytes

glucose transporter 5 (GLT5) intestinal transporter thought to be responsible for transport of fructose across the apical membrane of enterocytes

GnRH also known as LHRH

Golgi apparatus an organelle responsible for the modification of proteins prior to their release from a cell

gonads general term for structures which produce haploid gametes—ova or sperm

Glossary

growth hormone (GH) hormone released from the anterior pituitary gland which promotes growth through a number of different mechanisms

growth hormone inhibiting hormone (GHIH) also known as somatostatin, hypothalamic hormone which inhibits the release of GH from the anterior pituitary gland

growth hormone releasing hormone (GHRH) hypothalamic hormone which stimulates the release of GH from the anterior pituitary gland

G-proteins membrane proteins which link receptors to intracellular systems which produce second messengers

haemoglobin (Hb) oxygen transport pigment found in red blood cells

heart rate number of cardiac muscle contractions per minute

high density lipoprotein a lipoprotein involved in the transport of fats

homeostasis maintenance of a relatively constant internal environment

hormones chemical substances which initiate biological responses in cells

human chorionic gonadotrophin (hCG) hormone which maintains the endometrium in the first stages of pregnancy

hyperlipidaemia elevation in circulating triglyceride concentration above the normal range

hypernatraemia elevation in plasma sodium concentration from normal range

hypertrophy increase in size of a cell or organ

hyperventilation increase in the rate and depth of breathing

hyponatraemia reduction in plasma sodium concentration from normal range

hypophyseal stalk tissue containing neurons and blood vessels which links the hypothalamus and pituitary gland

hypothalamus region of the brain lying beneath the thalamus and involved in the regulation of a number of physiological processes (e.g. temperature regulation, eating and drinking behaviour, and regulation of the pituitary gland)

hypoxia general term for a lack of oxygen

ileum final section of the small intestine

inspiratory reserve volume (IRV) the volume of air that can be inspired over and above tidal volume

insulin hormone released from the pancreas which reduces plasma levels of glucose by stimulating its uptake into liver and muscle cells

insulin-like growth factors (IGFs) substances which mediate the effects of growth hormone

interatrial septum wall separating the left and right atria

internal elastic membrane layer of elastic fibres found in the tunica intima layer of arteries

interneurons neurons contained wholly within the central nervous system

interstitial fluid the layer of fluid in immediate contact with each and every cell in the body

interventricular septum wall separating the left and right ventricles

intracellular fluid (ICF) fluid which is found within cells, characterized by a large potassium concentration

intramural plexus part of the enteric nervous system consisting of the submucosal plexus and myenteric plexus

islets of Langerhans the endocrine portion of the pancreas producing insulin and glucagon

isometric contraction muscular contraction that results in an increase in force with no change in muscle length

isotonic contraction muscular contraction that results in an increase in force with a change in muscle length

jejunum second section of the small intestine

juxtaglomerular apparatus area consisting of macula densa and juxtaglomerular cells that is the site of renin production

kidneys main organs of the renal system that are involved in filtration of plasma and formation of urine

Korotkoff sounds sounds heard via a stethoscope applied to the brachial artery that can be used to determine systolic and diastolic blood pressure

lactation the process of milk production

lacteal lymphatic capillary found within intestinal villi

lamina propria layer of dense connective tissue found in the mucosal layer of the digestive system wall

large intestine Distal portion of gastrointestinal tract responsible for some absorption of water and electrolytes

larynx a structure containing the glottis and vocal cords

length–tension relationship relationship that describes the effect of muscle, or sarcomere, length on the production of tension

leptin hormone secreted from adipose tissue and involved in the regulation of long-term energy balance

leukocytes white blood cells involved in immune response

Leydig cells cells in the seminiferous tubules which produce testosterone

lipids fat-like molecules which perform a variety of roles in cells, including the formation of the cell membrane

liver large organ in the abdominal cavity with wide-ranging roles including protein synthesis, detoxification, and secretion of digestive enzymes

loop of Henle hairpin structure in a renal tubule which consists of the descending and ascending limbs

low density lipoprotein a lipoprotein involved in the transport of fats

lower oesophageal sphincter layer of smooth muscle at the base of the oesophagus

luteinizing hormone (LH) hormone released from the anterior pituitary gland which is responsible for gonadal function

luteinizing hormone releasing hormone (LHRH) also known as GnRH, hormone produced by the hypothalamus which stimulates the secretion of LH and FSH from the anterior pituitary gland

lysosome an organelle containing hydrolytic enzymes

major calyx formed by the fusion of numerous minor calyces; transports urine to the renal pelvis

mastication physical process involved in digestion of nutrients

mean arterial pressure average pressure within systemic arteries during a cardiac cycle

mechanoreceptors receptors which respond to and signal changes in pressure (e.g. touch receptors, pressure receptors, etc.)

medulla the region of the brainstem continuous with the spinal cord

menstrual cycle the cyclical changes which occur in the lining of the uterus (endometrium)

microglia phagocytic glial cells which remove debris from the nervous system

midbrain uppermost region of the brainstem

migrating myoelectrical complex wave of smooth muscle contraction throughout the stomach and intestines

Glossary

minor calyx area into which urine drains from the base of a renal pyramid before moving to a major calyx

mitochondria organelles responsible for energy transduction and the production of ATP in cells

mixed micelle formation of lipids found in the intestine prior to absorption

monogenic syndrome defect in a single gene that leads to a disorder

monosaccharide simple sugar containing five or six carbon atoms

motilin hormone secreted from the small intestine leading to increased gastric motility

motor endplate region of a skeletal muscle fibre membrane under the axon terminal of a motor neuron

motor neuron a myelinated nerve cell innervating a single skeletal muscle fibre

motor unit a motor neuron and the group of skeletal muscle fibres it innervates

mucosa outer layer of the digestive system wall consisting of epithelial cells, the lamina propria, and smooth muscle

multi-unit smooth muscle smooth muscle that does not act as a single unit but instead contracts in individual units

muscularis layer of the digestive system wall containing smooth muscle and the myenteric plexus

muscular tissue one of four fundamental tissue types whose principal function is movement

myelin sheath a lipid-rich layer formed by glial cells wrapped around the axon of a neuron—there are breaks in the myelin sheath called the nodes of Ranvier

myenteric plexus part of the intramural plexus found in the muscularis layer of the digestive system wall

myocardial infarction reduction in blood flow to the myocardium due to occlusion of the coronary arteries which leads to tissue necrosis

myocardium middle layer of the heart wall that consists of cardiac muscle

myofibril regular arrangement of actin and myosin within the cytoplasm of skeletal and cardiac muscle fibres

myometrium the muscular layer of the uterus

myosatellite cell progenitor cells found within skeletal muscle fibres that assist with muscle growth and repair

myosin key protein found within muscle fibres

myosin light chain kinase enzyme that phosphorylates myosin heads in smooth muscle fibres

Na^+/K^+-ATPase pump a pump located in cell membranes which transfers two K^+ ions into a cell in exchange for three Na^+ ions out of the cell

negative feedback a mechanism which contributes to the maintenance of homeostasis—disruption in a variable initiates responses which return the variable back to its set point

nephron functional unit of the kidney consisting of a renal corpuscle and a renal tubule

neural tissue one of four fundamental tissue types which forms the cells of the nervous system

neuropeptide Y neuropeptide secreted as part of orexigenic pathways

neuroendocrine control the release of chemical substances from axon terminals into plasma which then travel some distance to influence the activity of cells

neurons functional cells of the nervous system responsible for the generation and conduction of action potentials

neurotransmitters chemical substances released from axon terminals which influence electrical activity in other cells

nociceptors receptors which respond to tissue damage and noxious insult, also known as pain receptors

nodes of Ranvier breaks in the myelin sheath along the axons of some neurons

non-shivering thermogenesis production of heat by metabolic means when body temperature drops

noradrenaline a neurotransmitter used by the sympathetic nervous system, also a neurohormone released from the adrenal medulla

nucleic acids two types of molecule—DNA and RNA—which are involved in the storage and transmission of genetic information

nucleus the organelle which houses the chromosomes of a cell

obesity excessive accumulation of adipose tissue that poses a risk to health

oesophagus muscular tube connecting the mouth to the stomach

oestrogens general term for some of the female sex hormones

oligodendrocytes glial cells which form the myelin sheath of neurons in the central nervous system

oogonia cells which develop to form haploid ova

orexigenic neural pathways that lead to an increase in energy intake and/or a decrease in energy expenditure

organelle a general term for a variety of structures found within cells (e.g. mitochondria)

osteoarthritis mechanical degradation of joints and associated tissue that results in pain

osteoblasts bone cells which result in bone calcification

osteoclasts bone cells which result in the breakdown of bone

ovaries gonads of the female reproductive tract responsible for ensuring that an ovum is prepared for fertilization

oxygen–haemoglobin saturation curve the graphical representation of the relation between oxygen saturation of blood and the partial pressure of oxygen

oxyhaemoglobin haemoglobin that has been oxygenated

oxytocin hormone released from the posterior pituitary gland which causes uterine contractions during parturition

pacemaker potential gradual increase in membrane potential observed in cells within the sinoatrial node and some smooth muscle fibres that leads to spontaneous action potentials

pancreas organ with endocrine and exocrine functions

pancreatic polypeptide hormone secreted from the pancreas in response to food ingestion which regulates pancreatic function and has been implicated in the regulation of appetite

papillary muscles muscles that attach to heart valves and the heart wall whose contraction or relaxation leads to opening or closing of the valve

paracrine control endocrine control whereby chemical substances released from cells diffuse in interstitial fluid to nearby cells to influence their activity, sometimes called local hormones

parasympathetic ANS part of the autonomic nervous system concerned with rest and digestion (e.g. increase gut motility and secretion)

parathyroid glands endocrine glands associated with the thyroid gland which control plasma calcium levels

parathyroid hormone (PTH) hormone released from the parathyroid gland which increases plasma calcium levels

parietal cells cells within the stomach that secrete hydrochloric acid

parturition the process of childbirth

penis male reproductive organ responsible for delivery of semen to the female

Glossary

peptide YY hormone secreted from the ileum and colon that has been implicated in the regulation of appetite

pericardium fluid-filled sac consisting of two layers of connective tissue which surrounds the heart

perimysium layer of connective tissue surrounding skeletal muscle fascicles

peritubular capillaries capillaries surrounding renal tubules that are involved in the reabsorption of substances from renal filtrate

pharynx a structure common to both the digestive and respiratory systems, better known as the throat

phosphodiesterase enzyme responsible for the breakdown of cAMP

pia mater innermost layer of the meninges which adheres to the brain and spinal cord

pituitary gland endocrine gland located in the brain which is divided into two regions—anterior and posterior pituitary gland

plasma membrane a protein–lipid structure which forms the boundary of cells and organelles

platelets cell fragments found in blood which lead to haemostasis and blood clot formation

pleura membranes which cover both the lungs and the inside of the thoracic cavity

pneumocytes cells which form the alveoli—two types, type I and type II

pneumotaxic centre a collection of neurons who inhibit activity in inspiratory neurons

polysaccharide chain of monosaccharides formed by repeated dehydration synthesis reactions

pons region of the brainstem between the medulla and the midbrain

portal blood vessels blood vessels which link the hypothalamus and the anterior pituitary gland

posterior pituitary gland region of the pituitary gland which secretes the hormones vasopressin and oxytocin from neurons which have their cell bodies in the hypothalamus

pre-Botzinger complex region of the ventral respiratory group where inspiration is thought to be initiated

preload the effect of cardiac muscle fibre stretch on stroke volume; it is closely related to end-diastolic volume

pre-optic nucleus neurons in the hypothalamus which synthesize the hormone vasopressin to be released from axon terminals located in the posterior pituitary gland

progesterone a female sex hormone

prolactin hormone responsible for milk production in breast tissue

prolactin inhibiting factor (PIF) also known as dopamine, hypothalamic hormone which inhibits prolactin release from the anterior pituitary gland

proliferative phase the first half of the menstrual cycle when the endometrium is being restored to full functionality

pro-opiomelanocortin (POMC) polypeptide secreted as part of anorexigenic pathways

protein a macromolecule formed as a polymer of amino acids

protein kinases enzymes which phosphorylate substrates which, in the case of proteins, induces conformational changes and therefore activity in the protein

proximal tubule first section of a renal tubule in which tubular reabsorption takes place; consists of the proximal convoluted tubule and the proximal straight tubule

pulmonary circulation system of vessels that carry deoxygenated blood from the right ventricle to the lungs and return oxygenated blood to the left atria

pulmonary valve valve between the right ventricle and the pulmonary artery

Purkinje fibres specialized cells which propagate action potentials to the myocardium

pyloric region lower section of the stomach that ends at the pyloric sphincter

pyloric sphincter layer of smooth muscle at the base of the pyloric region

rapidly adapting receptors sensory receptors which respond to continued stimulation by significantly reducing the frequency of the action potentials they generate

receptors membrane or cytoplasmic protein molecules which neurotransmitters and hormones bind to initiate biological responses in cells

rectum storage site for waste products produced in the gastrointestinal system

refractory period the period of time immediately following the arrival of one stimulus to a neuron when it is either impossible or very difficult to generate a second action potential with a second stimulus

release hormones hormones released from the hypothalamus which promote the release of hormones from the anterior pituitary gland

release-inhibiting hormones hormones released from the hypothalamus which inhibit the release of hormones from the anterior pituitary gland

renal column area of renal cortex that projects between renal pyramids

renal corpuscle part of a nephron which consists of Bowman's capsule and a glomerulus and filters plasma before the filtrate is moved to the renal tubule

renal cortex outer portion of each kidney

renal fraction percentage of cardiac output delivered to the kidneys

renal medulla inner portion of each kidney containing renal pyramids and renal columns

renal pelvis formed by the fusion of several major calyces; transports urine to the ureter

renal pyramid triangular section of tissue found within the renal medulla which contains tubules that transport urine to a minor calyx

renal tubule part of a nephron where water and other substances are added to or reabsorbed from the filtrate from the renal corpuscle

renin hormone secreted from the juxtamedullary apparatus within the nephron which plays a central role in the renin–angiotensin–aldosterone system

repolarization the second phase of an action potential which sees the membrane potential reduce from the peak of the action potential back towards resting membrane potential

respiratory centre general term for brainstem neurons involved in the generation of respiratory rhythm

respiratory intercostal muscles muscles occupying the spaces between the ribs—contraction of these muscles contributes to both inspiration and expiration

respiratory muscle pump increases in venous pressure as a result of contraction of respiratory muscles which leads to an increase in venous return

resting membrane potential (RMP) potential difference (voltage) measured across the membrane of a neuron when it is at rest—it is of the order of 80mV, with the inside of the neuron being negative with respect to the outside

reticular formation a functional region of the brain, with areas in the brainstem and elsewhere which are involved in the regulation of wakefulness

rugae folds within the stomach that allow distension to occur

residual volume (RV) the volume of air left in the lungs following a maximal expiration

sarcomere contractile unit of skeletal and cardiac muscle

sarcoplasmic reticulum similar to smooth endoplasmic reticulum; it is found in muscle fibres

satiation the processes that lead to termination of energy intake

satiety the feeling of fullness that exists after termination of energy intake

secretin hormone released from small intestine and increases bicarbonate secretion

semen a substance containing sperm and a variety of other secretions released at ejaculation

Glossary

seminiferous tubules the site of sperm production in the testes

serosa innermost layer of the digestive system wall consisting of connective tissue

Sertoli cells cells in the seminiferous tubules where spermatogenesis occurs

shivering production of heat by asynchronous muscular contraction

single-unit smooth muscle smooth muscle that does not act as individual units but instead contracts as a single unit

sinoatrial node specialized group of cells in the right atrium that exhibit pacemaker potentials

skeletal muscle fibre multinucleated cylindrical cell found in skeletal muscle

skeletal muscle pump increase in venous pressure as a result of skeletal muscle contraction which leads to increased venous return

sleep apnoea disorder of sleep characterized by transient cessation of breathing

sliding-filament mechanism process of interaction between actin and myosin which leads to muscular contraction

slowly adapting receptors sensory receptors which respond to prolonged stimulation by maintaining their frequency of action potential production

small intestine main site of nutrient absorption, consisting of the duodenum, jejunum, and ileum

smooth muscle fibre mononucleated cell found in smooth muscle

sodium-linked glucose transporter 1 (SGLT-1) intestinal transporter responsible for the co-transport of sodium and glucose across the apical membrane of enterocytes

somatosensory refers to a whole range of sensory experiences originating from the body (e.g. temperature, pain, touch, etc.)

somatostatin see GHIH

special senses sensory information relating to taste, smell, vision, and hearing

sperm functional male gametes

spermatogenesis the process of producing fully mature sperm

spermatogonia cells from which sperm will be produced

sphincter of Oddi layer of smooth muscle that controls secretion of bile into the small intestine

sphygmomanometer apparatus used to measure blood pressure in the systemic circulation

spinal cord elongated neural tissue continuous with the medulla of the brain; it transmits action potentials between the periphery and the brain and also controls simple neural activity, i.e. spinal reflexes

spinal reflexes simple pre-programmed neural activity which is dealt with at the level of the spinal cord, although higher influences can voluntarily affect them

spirometry technique used to measure a range of lung volumes

steroid-binding protein plasma protein which transports steroid hormones within the plasma

steroid hormones a group of hormones derived from cholesterol which have a variety of actions within the body

stomach storage site for ingested nutrients and the site of some digestive processes

striated alternating bands of thick and thin proteins observed when examining skeletal or cardiac muscle

stroke volume volume of blood ejected from the heart in each cardiac muscle contraction

submucosa layer of the digestive system wall consisting of dense connective tissue; contains the submucosal plexus

submucosal plexus part of the intramural plexus found in the submucosal layer of the digestive system wall

supra-optic nucleus neurons in the hypothalamus which synthesize the hormone oxytocin that is released from axon terminals located in the posterior pituitary gland

surfactant a detergent-like substance produced by type II pneumocytes which reduces surface tension in the alveoli and prevents their collapse

susceptibility gene polymorphism in a gene that results in greater susceptibility to disease

sympathetic ANS part of the autonomic nervous system concerned with fear, flight, and fight responses (e.g. decreased gut activity, increased heart rate, and pupil dilation)

synapse physical gap separating the axon terminal of a neuron from another cell

systemic circulation system of vessels that carry oxygenated blood from the left ventricle to active tissue and return deoxygenated blood to the right atria

systole ejection phase of the cardiac cycle

systolic blood pressure pressure in the systemic circulation arteries at the end of systole

tendon fibrous tissue which connects skeletal muscle to bone

testes the male gonads responsible for the production of sperm

testosterone the principal male sex hormone

tetra-iodothyronine (T_4) a hormone released from the thyroid gland

thalamus region of the diencephalon which is responsible for the filtering of sensory information before it is relayed to higher regions of the brain

thermoreceptors receptors which monitor and respond to changes in both core temperature and peripheral temperature

thermoregulation process of heat loss or gain that leads to maintenance of core body temperature

thoracic cavity the cavity containing the heart, lungs, and various other structures (e.g. oesophagus)

threshold value of resting membrane potential which must be reached in order for an action potential to be generated—usually about 10–15mV above resting membrane potential

thyroid gland an endocrine structure in the neck which releases the hormones T_3 and T_4

thyroid stimulating hormone (TSH) hormone released from the anterior pituitary gland which stimulates the thyroid gland

thyrotrophin releasing hormone (TRH) hormone released from the hypothalamus which stimulates the release of TSH from the anterior pituitary gland

tidal volume (VC) the volume of air inspired or expired during normal quiet breathing

total peripheral resistance resistance to blood flow in the systemic circulation

trachea large-diameter tube which links the upper respiratory tract to the lower respiratory tract

transcellular fluid ECF which is not plasma and is found in a variety of locations (e.g. CSF, synovial fluid)

transverse tubules tubes filled with ECF extending from the sarcolemma into skeletal and cardiac muscle fibres

tricuspid valve valve between the right atria and the ventricle

tri-iodothyronine (T_3) hormone released from the thyroid gland

trophoblast the outer layer of the blastocyst

tropomyosin long protein found in skeletal and cardiac muscle that covers actin-binding sites and prevents the formation of cross-bridges

Glossary

troponin small protein found in skeletal and cardiac muscle fibres that is involved in the regulation of muscular contraction

tubular reabsorption process of reabsorption of solutes and water from the renal filtrate within the renal tubule

tubular secretion process of secretion of substances into the renal filtrate within the renal tubule

tunica externa outer layer of blood vessels, consisting largely of connective tissue, which provides stability to the vessel

tunica intima inner layer of blood vessels consisting largely of endothelial cells and connective tissue

tunica media middle layer of blood vessels consisting of smooth muscle fibres

upper oesophageal sphincter layer of skeletal muscle guarding the top of the oesophagus

ureters muscular tubes that transport urine from the kidney to the bladder

urethra tube extending from the bladder that transports urine for elimination

urine liquid excreted from the renal system which consists of water-soluble metabolic by-products

uterus muscular structure in the female pelvis which houses the developing fetus

vagina the opening of the female reproductive tract

varicocities swellings of post-ganglionic autonomic nerve fibres that release neurotransmitters

vas deferens a contractile tube in the testes into which the epididymi drain

vasoconstriction reduction in blood vessel diameter due to contraction of smooth muscle within the tunica media

vasodilation increase in blood vessel diameter due to relaxation of smooth muscle within the tunica media

vasopressin also known as arginine vasopressin or antidiuretic hormone, hormone secreted from the anterior pituitary gland which acts on the collecting ducts in the kidney to increase reabsorption of water

veins vessels that carry blood towards the heart

venous return volume of blood returned to the heart

ventral respiratory group region of the respiratory centre which contains neurons involved in both inspiration and expiration

ventricles chambers of the heart that eject oxygenated blood into the systemic circulation (left ventricle) and deoxygenated blood into the pulmonary circulation (right ventricle)

venules extensions of capillaries that carry blood towards the heart

very low density lipoprotein a lipoprotein involved in the transport of fats

villi projections of the mucosal layer of the digestive system wall that increase surface area

vital capacity (VC) maximum amount of air that can be exhaled following a maximum inspiration—the sum of IRV + VT + ERV

zona fasciculata the middle layer of the adrenal cortex which secretes glucocorticoids (e.g. cortisol)

zona glomerulosa the outermost layer of the adrenal cortex which secretes mineralocorticoids (e.g. aldosterone)

zona reticularis the innermost layer of the adrenal cortex which produces small amounts of the sex hormones

zygote the product of an ovum fertilized by a sperm which will develop into an embryo and then a fetus

Index

Index

Index